YOOL KIM, ELLIOT AXELBAND, ABBY DOLL, MEL EISMAN,
MYRON HURA, EDWARD G. KEATING, MARTIN C. LIBICKI, BRADLEY MARTIN,
MICHAEL E. MCMAHON, JERRY M. SOLLINGER, ERIN YORK,
MARK V. ARENA, IRV BLICKSTEIN, WILLIAM SHELTON

ACQUISITION OF SPACE SYSTEMS

PAST PROBLEMS AND FUTURE CHALLENGES

VOLUME 7

Prepared for the Office of the Secretary of Defense

NATIONAL DEFENSE RESEARCH INSTITUTE

For more information on this publication, visit www.rand.org/t/MG1171z7

Library of Congress Control Number: 2015933393

ISBN: 978-0-8330-8895-6

Published by the RAND Corporation, Santa Monica, Calif.
© Copyright 2015 RAND Corporation
RAND® is a registered trademark.

Cover image: United Launch Alliance

Support RAND
Make a tax-deductible charitable contribution at
www.rand.org/giving/contribute

www.rand.org

Preface

Space systems deliver critical capability to warfighters; thus, acquiring and deploying space systems in a timely and affordable manner is important to U.S. national security. However, many Department of Defense (DoD) space programs experienced large cost growth, schedule delays, and unanticipated technical problems for years, raising concerns about potential operational gaps in some critical space capabilities, because satellites were not being delivered as scheduled to replace the aging legacy systems in orbit. The difficulties faced during development of these systems may have been mostly resolved, because the systems have been delivered or are entering the production phase, but as DoD plans for the next-generation space systems in increasingly challenging fiscal and security environments, it is important to apply lessons learned from the past DoD space acquisition experience. RAND was asked to identify key factors that contributed to the difficulties in DoD space acquisition.

This report should interest policymakers concerned with military acquisition and related issues. It was sponsored by the Office of the Secretary of Defense (OSD) Performance Assessments and Root Cause Analysis (PARCA) office and conducted within the Acquisition and Technology Policy Center of the RAND National Defense Research Institute, a federally funded research and development center sponsored by the Office of the Secretary of Defense, the Joint Staff, the Unified Combatant Commands, the Navy, the Marine Corps, the defense agencies, and the defense Intelligence Community. Other RAND research sponsored by PARCA includes:

- Irv Blickstein et al., *Root Cause Analyses of Nunn-McCurdy Breaches, Volume 1: Zumwalt-Class Destroyer, Joint Strike Fighter, Longbow Apache, and Wideband Global Satellite,* Santa Monica, Calif.: RAND Corporation, MG-1171/1-OSD, 2011.
- Irv Blickstein et al., *Root Cause Analyses of Nunn-McCurdy Breaches, Volume 2: Excalibur Artillery Projectile and the Navy Enterprise Resource Planning Program, with an Approach to Analyzing Complexity and Risk,* Santa Monica, Calif.: RAND Corporation, MG-1171/2-OSD, 2012.

- Irv Blickstein et al., *Root Cause Analyses of Nunn-McCurdy Breaches, Volume 3: Joint Tactical Radio System, P-8A Poseidon, and Global Hawk Modifications*, Santa Monica, Calif.: RAND Corporation, MG-1171/3, 2013.
- Mark Arena et al., *Management Perspectives Pertaining to Root Cause Analyses of Nunn-McCurdy Breaches, Volume 4: Program Manager Tenure, Oversight of Acquisition Category II Programs, and Framing Assumptions.* Santa Monica, Calif.: RAND Corporation, MG1171/4, 2013.

For more information on the RAND Acquisition and Technology Policy Center, see http://www.rand.org/nsrd/ndri/centers/atp.html or contact the director (contact information is provided on the web page).

Contents

Preface iii
Figures and Tables vii
Summary ix
Acknowledgments xiii
Abbreviations xv

CHAPTER ONE
Introduction 1
Long-Standing Concerns about Space Acquisition 1
Overview of Past Literature 2
Purpose and Tasks 4
Research Approach and Scope 5
How the Report Is Organized 5

CHAPTER TWO
Review of Selected DoD Space Programs 7
Space-Based Infrared System 7
Global Positioning System IIF 12
Advanced Extremely High Frequency Program 15
Wideband Global SATCOM 19
Global Positioning System III 22
Summary 26

CHAPTER THREE
Effects of the 1990s' Space Acquisition Environmental Factors on the Programs 29
Shift in Requirements Increased Complexity 29
Acquisition Reform 32
Commercial Space Market 35
Summary 36

CHAPTER FOUR

Space Acquisition Challenges and Space Enterprise Management Issues............39
Space Acquisition Challenges...39
Space Enterprise Management Issues..44
Summary...53

CHAPTER FIVE

Recent Progress and Future Challenges in DoD Space Acquisition........55
Recent Progress in Space Acquisition..55
Future Challenges and Potential Implications for Future Space Acquisition Programs.....60
Lessons Learned from the 1990s 61
Summary...63

CHAPTER SIX

Conclusions..65

APPENDIXES

A. List of Interviews...67
**B. Projecting Defense Space System Budget Growth: Issues of Inflation Index
     Selection**...69

Bibliography...89

Figures and Tables

Figures

1.1. Sources of Space Systems Growth 3
2.1. SBIRS Program Cost Growth History 8
2.2. SBIRS Program Schedule Overrun History 11
2.3. History of GPS IIF Cost Growth 13
2.4. GPS IIF Program Schedule Overrun History 15
2.5. AEHF Program Cost Growth History 17
2.6. AEHF Program Schedule Delay History 19
2.7. History of the WGS Program Cost Growth 20
2.8. WGS Program Schedule Delay History 23
2.9. GPS III Cost Growth History 26
4.1. Long Gaps Between Orders and Unpredictable Buy Schedule 40
4.2. Vicious Circle Between Satellite High Reliability and Satellite Cost 44
B.1. Space Systems Cost Index Logic 73
B.2. Validation Process Example of the Accuracy of GDP Inflation Rate
Predictions 83

Tables

2.1. Sources of SBIRS Program Total Cost Growth (in Decreasing Order of
Percentage Contribution) 10
2.2. Sources of GPS IIF Program Total Cost Growth (in Decreasing Order of
Percentage Contribution) 14
2.3. Sources of AEHF Program Total Cost Growth (in Decreasing Order of
Percentage Contribution) 18
2.4. Sources of WGS Program Total Cost Growth (in Decreasing Order of
Percentage Contribution) 23
2.5. Sources of GPS Program Total Cost Growth (in Decreasing Order of
Percentage Contribution) 27
2.6. Performance of Selected DoD Space Programs 28
3.1. Increased Tactical Mission Requirements 31
4.1. Examples of Space and Launch Segment Synchronization Challenges 45

4.2. Gap Between Fielding of Satellites and Their Ground Control and User Segments...47
4.3. Factors That Contributed to Schedule Pressures in DoD Space Programs 51
4.4. Shift in Risk Posture from Early Space Programs to Current Space Programs.....52
5.1. Recent Efficiency and Cost-Saving Initiatives in Selected DoD Space Systems in Production..57
A.1. Interviews Conducted in Support of the Research 67
B.1. Government Organizations Visited..73
B.2. Defense and Producer Price Index Deflators Related to Defense 76
B.3. Inflation-Based Differences in WGS Block II Satellite Expected Unit Prices ...79

Summary

Background and Purpose

Space systems deliver critical capability to warfighters; thus, acquiring and deploying these systems in a timely and affordable manner is important to U.S. national security. However, many Department of Defense (DoD) space programs have experienced large cost growth, schedule delays, and unanticipated technical problems. During this period of difficulties, concerns about potential operational gaps in some critical space capabilities were raised, because satellites were not being delivered as scheduled to replace the aging systems in orbit. The difficulties faced during development of these systems seem to have been mostly resolved, because the systems have been delivered or are entering the production phase. However, as DoD plans for the next-generation space systems in increasingly challenging fiscal and security environments, it is important to apply lessons learned from past DoD space acquisition experience. RAND was asked to identify key factors that contributed to the difficulties in military space system acquisition, specifically, cost growth, schedule delays, and technical issues.

Approach

We focused on identifying enterprise-level systemic issues that contributed to the space acquisition difficulties. Our analysis comprises four tasks:

- Analyze the performance of selected DoD space programs in terms of cost growth, schedule delays, and satellite on-orbit performance (Space-Based Infrared System [SBIRS], Global Positioning System IIF [GPS IIF], Advanced Extremely High Frequency [AEHF] program, Wideband Global SATCOM [WGS], Global Positioning System III [GPS III]).
- Identify enterprise-level systemic issues and key factors that contributed to cost growth, schedule overruns, and technical problems in selected DoD space programs.
- Characterize the current status of the selected DoD space programs.

- Identify future acquisition challenges that the next-generation space systems might face.

The methodology we used to carry out the research had three components. First, we examined relevant data in the program status reports that the program office provides: the Selected Acquisition Reports and the Defense Acquisition Executive Summary reports. Second, we reviewed literature on the relevant topics. Third, we interviewed acquisition officials and contractors to gain insights into the context within which decisions were made and the key factors that contributed to space acquisition difficulties.

Key Factors Contributing to Space Acquisition Difficulties

Four of the five selected DoD space programs we examined (SBIRS, GPS IIF, AEHF, and WGS) experienced major cost growth and schedule delays. The fifth program, GPS III, has seen some moderate cost growth, but it may be premature to judge its overall program performance. The four programs experienced major cost growth and schedule delays arising from difficulties in technology development, engineering, manufacturing, integration, parts quality, and obsolescence, which led to costly redesign, rework, and additional testing. These programs had implemented a high-risk acquisition approach that contributed to these difficulties and inefficiencies. The high-risk acquisition approach was characterized by the following three types of risk:

1. high requirements risks: midstream changes in requirements and complex and ambitious requirements arising from multiple missions of equal priority on a single platform
2. high technical risks: introduction of immature technologies, inadequate testing and systems engineering, and overoptimistic assumptions about applicability of commercial practices and standards for military space systems
3. high programmatic risks: accelerating program schedules, changes in procurement quantities, and inefficient buying practices that caused long production gaps.

We found that the external environmental factors (the geopolitical environment, fiscal environment, etc.) that space acquisition faced in the 1990s, when many of the DoD space programs we reviewed were being initiated, played a key role in the programs' decisions to implement such a high-risk acquisition approach. The end of the Cold War profoundly changed U.S. national security strategy and significantly reduced the defense budget in the 1990s when many current space programs started. Such changes led to new requirements for increased tactical support from space programs, a reduction in both the government and contractor workforce, and acquisition reform that focused on cost and shifted to commercial standards, products, and practices. These policy and strategy changes were implemented by introducing high risks into the programs without fully understanding the potential consequences. Such a high-risk acquisition approach was not appropriate for space systems, given their inherently challenging nature characterized by a limited industrial base, stringent standards, low volume and high unit cost, high technological complexity, and demand for high reliability.

We found several enterprise-level systemic issues that further contributed to the adverse outcome of the programs. First, the long mission life demanded from military space systems (partly driven by the high satellite and launch costs and inability to cost-effectively repair space systems hardware once on-orbit), although desirable in some ways, contributed to the long development cycles. Long development cycles create their own risks by generating more opportunities for requirements changes, funding reductions, obsolescence, or unanticipated technical problems. They also foster the temptation to use the latest but immature state-of-the-art technology. Second, lack of synchronization between all the segments in the space enterprise (satellite, launcher, ground control, and user equipment) contributed to schedule delays and unanticipated changes in the programs, such as changes in the launch vehicle. Changes in any one segment often required changes in the other segments with cost and schedule implications. Disparate management of the segments may also have contributed to the difficulties in synchronization. Third, poor constellation management characterized by multiple changes in the constellation size and mix during the life of the program, early termination of production of legacy systems, and limited ability to address unanticipated gaps in constellation created production gaps and schedule pressures on programs. Production gaps contributed to cost growth because of inefficient buying practices, and schedule pressures fostered a risky path to acquisition. Finally, the failure-intolerant risk posture combined with the goal of optimizing each satellite's utility (e.g., maximum use of satellite service life, using the "test" article as an operational system) allowed very little room for error for these long, complex development programs. There was minimal margin (schedule, technical, and cost) to deal with unanticipated problems, making space acquisition susceptible to cost growth and schedule overruns.

Implications for Future Space Acquisition

In recent years, progress has been made in space acquisitions. Four of five programs we examined are now transitioning into the production phase, and satellites are being launched. In part for this reason and in part because of recent efficiency and cost-saving initiatives, costs and schedules appear to be under better control.

However, future space acquisition faces a period of significantly reduced budgets, similar to that of the 1990s, which fostered a high-risk acquisition approach for space systems. Emerging threats are again shifting requirements and architectures for the space enterprise driven by the need for increased resilience. The drive for resilience is adding pressure and complexity to space acquisition. Further, as in the 1990s, austere budgets are likely to reduce acquisition and technical expertise. They also invite the reintroduction of themes (e.g., leveraging commercial systems, standards, and streamlining acquisition processes) prominent in the 1990s acquisition reform era that contributed to significant difficulties in the programs.

For development of future programs, the application of acquisition efficiency initiatives and alternative space architectures requires a careful assessment based on the degree of risk and the tolerance for failure associated with any particular program. Moreover, although alternative architectures, such as disaggregated architectures, could alleviate certain risk and complexity factors in space acquisition, these approaches would introduce other new risk and complexity factors, such as increasing complexity in synchronization and constellation management because of diversity in the architecture. The overarching conclusion is that there is no "silver bullet" to fix space acquisition difficulties. All realistic acquisition approaches require tradeoffs and the assumption of some risks. Comprehensive analyses to inform such tradeoffs (including tradeoffs at the enterprise level) are needed for a robust acquisition approach.

Acknowledgments

The research presented in this report benefited tremendously from discussions with many individuals in the Office of the Secretary of Defense, the Air Force, the Joint Staff, and the space industry involved in space acquisition. We would like to thank our sponsor in OSD, Mr. Gary Bliss, Director, Performance Assessments and Root Cause Analysis in the Office of the Assistant Secretary of Defense for Acquisition for his support and guidance. We thank Charles Beames, Principal Deputy, Space and Intelligence Office in the Office of the Assistant Secretary of Defense for Acquisition, and his staff. We also thank General William Shelton, Commander, Air Force Space Command, and his leadership team; Lieutenant General Ellen Pawlikowski, Commander, Space and Missile Systems Center, and her leadership team; Lieutenant General Charles R. Davis and Major General Robert McMurry in the Office of the Assistant Secretary of the Air Force for Acquisition and their staffs; and Richard McKinney, Deputy Under Secretary of the Air Force (Space), and his staff. Many other individuals shared their insights and provided information for our study. For brevity, we provide the names of their organizations. They include the OSD Cost Assessment and Program Evaluation, Joint Staff Force Structure, Resource, and Assessment Directorate, Headquarters Air Force Operations, Plans, and Requirements Directorate, Space Development and Test Directorate in Space and Missile Systems Center, Headquarters Air Force Space Command Requirements Directorate, Headquarters Air Force Space Command Strategic Plans, Programs, and Analysis Directorate, the Aerospace Corporation, Boeing Space and Intelligence Systems, Lockheed Martin Space Systems Company, Northrop Grumman Corporation, and McKinsey & Company. The report also benefited from the insightful comments and suggestions provided by our RAND colleagues Bernard Fox and Lara Schmidt.

Abbreviations

ACAT	Acquisition Category
AEHF	Advanced Extremely High Frequency program
AFCAA	Air Force Cost Analysis Agency
AFSPC	Air Force Space Command
APB	acquisition program baseline
APUC	average procurement unit cost
ASIC	application-specific integrated circuit
AT&L	Acquisition, Technology, and Logistics
ATP	authority to proceed
BEA	Bureau of Economic Analysis
BLS	Bureau of Labor Statistics
BY	base year
CAAG	Acquisition, Technology, and Logistics
CAPE	Cost Assessment and Program Evaluation
CBO	Congressional Budget Office
CLS	contractor logistics support
DAES	Defense Acquisition Executive Summary
DCMA	Defense Contract Management Agency
DoD	U.S. Department of Defense
DOT&E	Director, Operational Test and Evaluation
DSB	Defense Science Board
DSCS	Defense Satellite Communication System
DSP	Defense Support Program

EAC	estimate at completion
EELV	Evolved Expendable Launch Vehicle
EMD	engineering and manufacturing development
EMI	electromagnetic interference
FAB-T	Family of Advanced Beyond Line-of-Sight Terminals
FFP	firm fixed price
FPGA	field-programmable gate array
FPRA	Forward Pricing Rate Agreement
FPRP	Forward Pricing Rate Proposal
FY	fiscal year
GAO	U.S. Government Accountability Office
GDP	Gross Domestic Product
GEO	geosynchronous or geostationary orbit
GNST	GPS III Non-Flight Satellite Testbed
GPS	Global Positioning System
HEO	highly elliptical orbit
ICBM	intercontinental ballistic missile
ICE	independent cost estimate
IMU	inertial measurement unit
IOC	initial operational capability
IR	infrared
KPP	key performance parameter
LM	Lockheed Martin
Mbps	megabits per second
MDAP	Major Defense Acquisition Program
MILSATCOM	military satellite communications
MOU	memorandum of understanding
NASA	National Aeronautics and Space Administration
NRE	nonrecurring engineering

NRO	National Reconnaissance Office
NSA	National Security Agency
O&S	operations and support
OCS	Operational Control Segment
OCX	Operational Control Segment (GPS III)
ODC	Other Direct Charges
ODNI/CA	Office of the Director of National Intelligence/Cost Analysis
OMB	Office of Management and Budget
OSD	Office of the Secretary of Defense
OUSD	Office of the Under Secretary of Defense
OUSD (A&T)	Office of the Under Secretary of Defense for Acquisition and Technology
OUSD (AT&L)	Office of the Under Secretary of Defense for Acquisition, Technology, and Logistics
OUSD (C)	Office of the Under Secretary of Defense, Comptroller
PARCA	Performance Assessments and Root Cause Analysis
PAUC	program acquisition unit cost
PM	program manager
PNT	Positioning, Navigation, and Timing
PPI	Producer Price Index
RDT&E	research, development, testing, and evaluation
RWA	reaction wheel assembly
SAF	Secretary of the Air Force
SAF/AQS	Under Secretary of the Air Force for Acquisition
SAF/FMC	Under Secretary of the Air Force for Cost and Economics
SAR	Selected Acquisition Report
SATCOM	satellite communications
SBIRS	Space-Based Infrared System
SDD	system development and demonstration

SPO	System Program Office
SV	space vehicle
TRL	technology readiness level
TSAT	Transformational Communications System
TSPR	Total System Performance Responsibility
TT&C	telemetry, tracking, and command
TY	then-year
UAV	unmanned aerial vehicle
WGS	Wideband Global SATCOM
XDR	extended data rate

CHAPTER ONE

Introduction

Long-Standing Concerns about Space Acquisition

Space systems deliver critical capability to warfighters: early warning of potential missile launches, assured communications capability, key intelligence, and timing and location information. Thus, acquiring and deploying space systems in a timely and affordable manner are important to U.S. national security. However, for years, Department of Defense (DoD) space programs suffered large cost growth, schedule delays, and unanticipated technical problems. For example, cost growth for satellite systems (measured over the five years after Milestone B[1]) exceeded cost growth for all other categories of defense systems.[2] During this period of difficulties, concerns were raised about potential operational gaps in some critical space capabilities, because satellites were not being delivered as scheduled to replace the aging systems in orbit.

The difficulties faced during development of these systems may be mostly resolved, because the systems have been delivered or are entering the production phase. At the moment, Global Positioning System III (GPS III) is the only satellite program still in development (although it is nearing the end of this phase). For the other satellite programs, several studies and technology demonstrations that provide points of departure for the development of follow-on systems have been either completed or are ongoing. As DoD plans for the next-generation space systems in increasingly challenging fiscal and security environments, it is important to gain a better understanding of the key factors and systemic issues that contributed to the difficulties in past DoD space acquisition.

---

[1]   Milestone B is a point at which an acquisition program enters the engineering and manufacturing development (EMD) phase.

[2]   Younossi et al., 2007. The median values of the program budget cost growth factors of DoD space Major Defense Acquisition Programs (MDAPs) were compared with DoD MDAPs from other sectors (e.g., military aircraft and Navy ships) based on a normalized cost-assessment approach using annual program Selected Acquisition Report (SAR) program budget cost, procurement quantity, and other reported data.

Overview of Past Literature

Because of the DoD space acquisition's prolonged difficulties, the literature over the past decade has examined possible causes of these difficulties. Most notable is the 2003 Defense Science Board (DSB) study (also know as the Young study), which was tasked to identify systemic issues in national security space programs.[3] It found the following as the "basic reasons" for the cost growth and schedule delays in space programs:

- a shift to a cost-focus from a mission-focus in managing space development programs
- unrealistic cost estimates
- undisciplined definition and uncontrolled growth in system requirements
- eroded government capabilities to lead and manage the space acquisition process
- failure to implement proven management and engineering practices by industry.

The U.S. Government Accountability Office (GAO) has repeatedly examined space acquisition programs over the past decade and has identified the following causes for space acquisition problems[4]:

- DoD starts more weapon programs than it can afford.
- DoD starts its programs too early, before it has the assurance that the capabilities it is pursuing can be achieved within available resources and time constraints.
- Programs have historically attempted to satisfy all requirements in a single step.
- Programs began in the late 1990s when government oversight was reduced and key decisionmaking responsibility was shifted to contractors.

GAO further identified the following negative influences that can cause programs to fail:

- optimistic cost and schedule estimates
- immature technology
- poorly understood software needs
- unstable requirements and funding
- inadequate contracting strategy
- inadequate contractor oversight
- lack of consideration of alternatives.

More recently, a more quantitative analysis is presented in the 2013 Defense Acquisition Performance Report.[5] This report stated that cost growth for space sys-

---

[3]   Office of the Under Secretary of Defense, Acquisition, Technology, and Logistics (OUSD [AT&L]), 2003.

[4]   GAO, 2012.

[5]   OUSD (AT&L), 2013b.

tems stemmed from four types of proximate causes, illustrated in Figure 1.1: contractor execution errors, work content changes, technology development difficulty, and integration difficulty. Note that no one factor was dominant. According to our interviews with personnel from the organization that conducted this study, the systemic issues that led to these difficulties were no surprise, in that they were mostly consistent with previous studies' findings: inadequate supplier management, increasing complexity, inadequate systems engineering, additional testing without fully understanding its costs and benefits, and an inadequate technology assessment process.

Other literature, including prior RAND studies, revealed similar findings, identifying issues related to requirements and funding instability, immature technologies entering acquisition, the complexity of space programs, poor program management practices, acquisition reform in the 1990s, leadership and workforce management issues, underestimation of costs, poor acquisition strategies, and so forth as contributing factors to cost growth, schedule delays, and technical difficulties.

In our literature review, we have not found any study that showed a causal relationship between these identified "causes" and poor program outcome. Indeed, a causal analysis would be extremely difficult to conduct, because it may require setting up an

Figure 1.1
Sources of Space Systems Growth

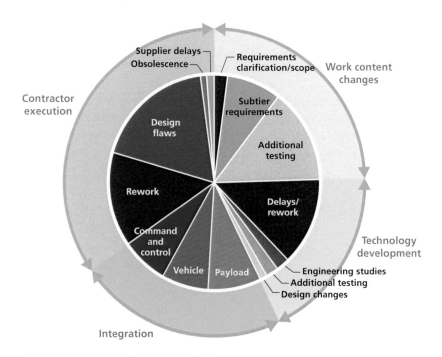

SOURCE: OUSD (AT&L), 2013b, Table 2-4.
RAND *MG1171/7-1.1*

experimental acquisition program.[6] Lack of documented information on the rationale for the key decisions made throughout a program's life further limits the ability to conduct a causal analysis. Further, it may be very difficult for decisionmakers to prioritize which "causes" of space acquisition difficulties should be addressed, because there are too many of them, and many of the "causes" may be tactical ones that address specific issues and thereby have a very limited effect on fixing higher-level space acquisition issues.

Purpose and Tasks

RAND was asked to determine key contributing factors and systemic issues that might explain the difficulties in military space system acquisition, specifically, cost growth, schedule delays, and technical issues. We did not conduct a causal analysis and did not attempt to identify all causes of acquisition difficulties. Rather, we focused on identifying enterprise-level systemic issues that contributed to the space acquisition difficulties to identify strategic ways to improve space acquisition.

Our analysis aims to accomplish four tasks:

Task 1: Analyze the performance of selected DoD space programs in terms of cost growth, schedule delays, and satellite on-orbit performance (Chapter Two).
Task 2: Identify enterprise-level systemic issues and key factors that contributed to cost growth, schedule overrun, and technical problems in space acquisition (Chapters Three and Four).
Task 3: Characterize the current status of the selected DoD space programs (Chapter Five).
Task 4: Identify future acquisition challenges that the next-generation space systems might face (Chapter Five).

Our sponsor also asked RAND to conduct separate research on defense space systems inflation rates and their effects on space systems costs, in addition to identifying key factors that contribute to defense space acquisition difficulties. Appendix B documents this separate research.

---

[6]   According to Donald Rubin (regarded by many as the leader in causal analysis): "For obtaining causal inferences that are objective, and therefore have the best chance of revealing scientific truths, carefully designed and executed randomized experiments are generally considered to be the gold standard. Observational studies, in contrast, are generally fraught with problems that compromise any claim for objectivity of the resulting causal inferences" (Rubin, 2008, p. 808).

Research Approach and Scope

We focused our analysis on the following five major DoD space systems (and associated ground control and user equipment, where applicable), taking them to be representative of the DoD space sector:[7]

- Space-Based Infrared System (SBIRS)
- Global Positioning System IIF (GPS IIF)
- Advanced Extremely High Frequency (AEHF)
- Wideband Global SATCOM (WGS)
- Global Positioning System III (GPS III)

    By contrast we have excluded:

- DoD satellite programs that are not Acquisition Category (ACAT) 1
- MUOS
- DoD systems that are not satellites, such as the Evolved Expendable Launch Vehicle (EELV) as well as Space Fence (even though they are technically in the space portfolio).

    The methodology we used to carry out the research had three components. First, we examined relevant data in the program status reports that the program office provides—SARs and the Defense Acquisition Executive Summary (DAES) reports. Second, we reviewed literature on this topic. Third, we interviewed acquisition officials and contractors to gain insights into the context within which decisions were made and potential systemic issues in space acquisition (see Appendix A).

How the Report Is Organized

The next chapter provides background information on the five selected DoD space programs and an analysis of each one's cost growth, schedule delays, and technical problems. It further discusses program decisions and attributes that contributed to such acquisition difficulties. Chapters Three and Four then discuss systemic issues and key contributing factors that influenced or shaped the program decisions and attributes. Specifically, Chapter Three examines key contributing factors that stemmed from the acquisition environment, and Chapter Four examines those related to challenges inherent in DoD space acquisition and space enterprise management issues. Then, in Chapter Five, we discuss the current status of space acquisition, including recent progress

---

[7]   The Air Force oversees DoD space programs with the exception of Mobile User Objective System (MUOS), which the Navy oversees.

in space acquisition and future challenges. Chapter Six contains a summary and suggestions for focus areas for improvements in space acquisition. Appendix A lists the organizations interviewed in support of the research, and Appendix B documents the space inflation index research.

CHAPTER TWO

Review of Selected DoD Space Programs

In this chapter, we analyze the performance of the five selected DoD space programs (SBIRS, GPS IIF, AEHF, WGS, and GPS III) in terms of cost growth, schedule delays, and satellite on-orbit performance using relevant data provided in SARs, the DAES, and open sources. We describe the program history and identify key program attributes and decisions that contributed to acquisition difficulties. Chapters Three and Four examine the systemic issues in space acquisition that influenced and shaped such adverse program attributes and decisions.

Space-Based Infrared System

SBIRS is the follow-on of the Defense Support Program (DSP) satellites that were designed to support one primary mission: missile warning, which they have effectively supported since 1971.[1] Unlike DSP, the SBIRS satellites were required to support four missions of equal priority: missile warning, missile defense, enhanced technical intelligence, and battle space characterization. SBIRS began as an EMD program in 1996 with immature technologies, complex requirements, and unrealistic cost estimates.[2] The original $2.3 billion contract awarded to Lockheed Martin was for a ground segment, two highly elliptical orbit (HEO) sensor payloads, and five geostationary (GEO) satellites.[3] Lockheed's proposal called for all-new sensors for the HEO payloads (scanner sensor) and GEO satellites (scanner and starer sensors[4]) rather than satellite designs that involved a modified heritage DSP payload.[5] Additionally, the SBIRS flight soft-

---

[1] As scientists and engineers learned from DSP satellites in orbit, they developed capabilities to support three other missions: technical intelligence, battlespace awareness, and theater missile warning.

[2] OUSD (AT&L), 2003; Younossi et al., 2008. The 2002 SBIRS Independent Review Team found two root causes: "program too immature to enter system design and development" and "system requirements decomposition and flow down not well understood as program continued to evolve" (Hura et al., 2007).

[3] All subsequent cost figures in this report are in fiscal year (FY) 2013 dollars.

[4] The scanning sensor detects a missile launch and then cues the staring sensor to forecast an incoming trajectory.

[5] Younossi, 2008.

ware was an internal research and development product that Lockheed Martin had used for its generic A2100 bus.[6] Despite such complexity and immature technologies, the projected cost of SBIRS satellites proposed in 1996 at $2.3 billion in FY 2013 dollars was much less than that of the much simpler DSP satellites.[7] This all suggests that the SBIRS EMD program was substantially underfunded from the beginning.

Figure 2.1 lays out a cost history for the SBIRS program. The figure shows the total program manager's (PM's) program cost estimate at completion (EAC) reported in the SARs since the EMD award. Eight SARs were released between June 2002 and December 2006, because multiple program rebaselining and recertification resulting from Nunn-McCurdy breaches required midyear SARs.[8] There was no SAR submission for December 2000 and December 2008. From the $2.3 billion at the November 1996 EMD award to December 2001, the cost EACs grew to $5 billion even with the reduction in the number of GEO satellites being acquired from five to two. The number of GEO satellites being acquired in the EMD contract was first reduced from five to three in 1998 when the fourth and fifth GEO satellites (GEO-4 and GEO-5) moved from research, development, testing, and evaluation (RDT&E) to the procure-

Figure 2.1
SBIRS Program Cost Growth History

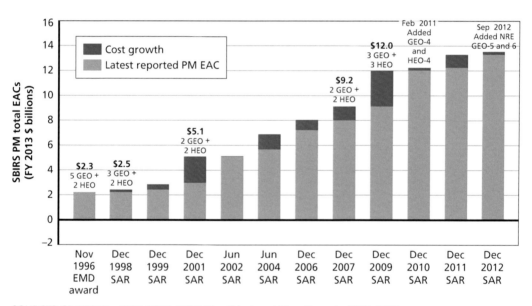

SOURCES: SBIRS SARs, 1996–2012; SBIRS Monthly Acquisition Reports, 2009–2012.
RAND MG1171/7-2.1

---

[6]   Space and Missile Systems Center (SMC), 2013b.

[7]   Hura et al., 2007.

[8]   A Nunn-McCurdy breach occurs when the procurement acquisition unit cost (PAUC) or average procurement unit cost (APUC) exceeds certain percentages.

ment budget. In 1999, the program office decided to defer the third GEO satellite (GEO-3) as part of a follow-on GEO block buy for GEO-3 through GEO-5, but the PM's EAC resulting from this scope change was not reported in the SAR until December 2001. The scope was further reduced in 1999 when a portion of the ground segment work was also deferred to procurement.

By December 2006, the estimated cost to completion grew to roughly $8 billion, and the program suffered four Nunn-McCurdy unit cost breaches, two of which required recertification. Most of the cost growth that occurred through 2006 was attributed to technical difficulties. The contractor encountered several major technical issues during the early years of the program, including the following:

- adding a 12-ft sunshade for all GEO satellites to meet performance requirements
- sensor issues related to the design, fabrication, and manufacturing, which caused problems associated with the sensor chip assembly needed to mount the sensor detector arrays
- pointing and control assembly software development difficulties
- GEO signal-processing software development
- GEO satellite parts obsolescence
- HEO-1 payload electromagnetic interference (EMI)
- HEO-1 payload software qualification.

Resolving these major issues and other technical problems required increasing mission assurance efforts (including additional component and system testing and ensuring part pedigrees), fixing hardware and software problems, and addressing alerts to problems in analogous parts (e.g., reaction wheel assembly from the same manufacturer), design, and engineering and manufacturing processes used in other programs. The program also faced much programmatic turbulence during this period, which contributed to inefficiencies in acquisition and program management and execution. For example, between 1998 and 2005, the program was rebaselined seven times.[9]

By December 2007, the program cost had grown to $9.2 billion for two GEO satellites and two HEO payloads—an increase of about $7 billion since the original EMD award. The cost growth after 2007 was primarily attributed to such programmatic factors as government scope changes, dominated by adding more satellites. A third GEO satellite (GEO-3) and a third HEO payload (HEO-3) were added in 2009, increasing the program EAC to $12 billion. In 2011, contracts for a fourth GEO satellite (GEO-4) and a fourth HEO payload (HEO-4) were awarded, increasing the EAC to about $13.3 billion. During the 2007 and 2011 time frame, the program scope grew to include other efforts, such as additional contractor logistics support (CLS), sustain-

---

[9]   Hura et al., 2007.

ment tasks, combined task force efforts, and a GEO testbed upgrade. The scope of the ground segment development also increased.

The cost growth resulting from technical issues moderated after 2008, with satellite development winding down and production beginning. However, after the December 2012 SAR was released, GAO reported in March 2013 that GEO-3 and GEO-4 were expected to be delayed by 14 months, partly as a result of technical challenges, parts obsolescence, and test failures, contributing to about $438 million in cost overruns.[10] The long production break between HEO-1 and HEO-3 and -4 and GEO-1 and GEO-3 and -4 (and the long development schedule of GEO-1 and HEO-1) caused significant parts obsolescence issues and introduced some additional nonrecurring engineering (NRE) efforts.

In 2012, the acquisition strategy for SBIRS GEO-5 and GEO-6 satellites was approved for $1.9 billion as part of DoD's Efficiency Space Procurement initiative, and the Air Force awarded a contract to Lockheed Martin in 2012 and again in 2013 for NRE and advanced procurement of long-lead parts. In June 2014, the Air Force awarded a fixed price contract to Lockheed Martin to complete production of these two satellites.[11] These satellites are expected to be delivered in the 2019–2020 time frame.[12]

Table 2.1 lists technical and programmatic cost-growth sources (e.g., government scope changes) that contributed to the total program cost growth of $11.3 billion. Within each category (technical or programmatic), the individual cost growth sources are listed in decreasing order of level of contribution (i.e., from the highest to the lowest

Table 2.1
Sources of SBIRS Program Total Cost Growth (in Decreasing Order of Percentage Contribution)

Technical Cost-Growth Source	Programmatic Cost-Growth Source
GEO Pointing and Control Assembly, flight, and signal processing software issues	Added follow-on buy of four GEO satellites and two HEO payloads
Added and integrated 12-ft sunshade to all GEO satellites	Added CLS, sustainment, and other tasks
HEO payload EMI issues	Increased scope of ground segment
Sensor design and parts fabrication issues (e.g., detector arrays)	Added combined task force and GEO-tested upgrade
HEO payload software qualification issues	
GEO parts obsolescence	

SOURCES: SBIRS SARs, 1996–2012; SBIRS Monthly Acquisition Reports, 2009–2012.

---

[10] GAO, 2013a, p. 120.

[11] Lockheed Martin, 2014.

[12] SBIRS SAR, 2012. On September 19, 2013, Lockheed Martin received a long-lead parts contract for SBIRS GEO-5 and -6 for $42 million (Space News Staff, 2013).

cost-growth source). We found that the amount of total cost growth attributed to technical issues is about the same as that attributed to programmatic factors.

The technical and programmatic difficulties that the SBIRS program encountered also affected the program schedule. Figure 2.2 shows a history of program schedule overruns based on the delivery dates of the first two HEO payloads and first two GEO satellites reported in SARs. The bar heights refer to the estimated number of years from EMD authority to proceed (ATP) to satellite or payload delivery. The solid bars are based on planned delivery dates, and the cross-hatched bars are based on the actual delivery dates. The planned or actual delivery dates are shown at the top of the bars. The HEO payloads were delivered about three years late. The first GEO satellite was delivered in March 2011, about nine years later than the originally planned delivery date of June 2002.

Today, both SBIRS HEO-1 and SBIRS HEO-2 are in service, and SBIRS GEO-1 and GEO-2 were launched in May 2011 and March 2013, respectively. SBIRS GEO-1 was declared operational in May 2013 after two years of on-orbit testing. The lengthy on-orbit testing was attributed to an onboard communications issue.[13] SBIRS GEO-2 received Air Force Space Command (AFSPC) operational acceptance in November 2013.[14]

Figure 2.2
SBIRS Program Schedule Overrun History

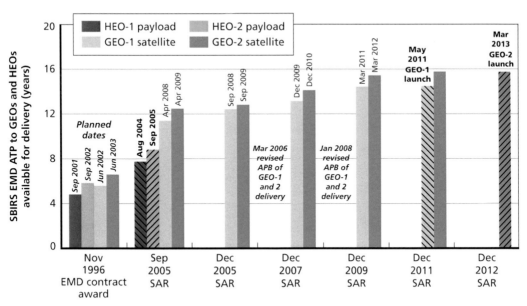

SOURCES: SBIRS Selected Acquisition Reports, 1996–2012.
RAND MG1171/7-2.2

---

[13] "Two Years Later, SBIRS Geo-1 Finally Declared Operational," 2013.

[14] Lockheed-Martin, 2013b.

Global Positioning System IIF

The GPS constellation provides valuable positioning, navigation, and timing (PNT) services to both civilian and military users. With a growing number of users and increasing demand for precision, GPS satellites that are near the end of their service lives are being replaced by improved satellites; ground control stations and user equipment are also being upgraded over time.

The GPS family evolved from 1974 beginning with GPS I satellites, followed by GPS II, GPS IIA, and GPS IIR satellites. In 1996, the GPS IIF contract for six satellites was awarded to Rockwell (now Boeing) for $550 million, with separately priced options written for two additional blocks, one of 15 satellites and the other of 12 satellites, all totaling 33 satellites.[15]

Three years after the original contract award, GPS IIF's requirements changed significantly following a presidential decision in 1999 to modernize GPS, to provide its services to all civilian users, and to introduce new military and civil codes.[16] These changes entailed fielding a new civil code (L2C) and a new military code (M-code). The requirements for those signals were added to the existing GPS IIR and IIF programs (eight GPS Block IIR-M space vehicles [SVs] with M-code plus L2C civil signal and all 12 GPS Block IIF SVs with M-code plus L2C and L5 civil signals).[17] However, it took three years for contract modifications to reflect the presidential decision. In the interim, Boeing was directed to conserve resources and mitigate potential rework. A new single acquisition master plan for GPS IIF was not developed until 2001,[18] and Boeing started the modernization activities under an Undefinitized Contract Action in 2002.

No GPS IIF satellites had been delivered under the prior contract by the time modernization was introduced. At this time, Boeing was completing subsystems for final satellite assembly for the first three IIF satellites. The program plan was revised to reflect the modernization requirements in May 2002 and was followed by a contract modification, which required retrofitting the first three IIF satellites with the additional signals and completing the design and production of the next three satellites. Partially built satellites had to be disassembled to recycle those components for new satellites.

One year later, the program plan was revised again with an amended acquisition program baseline (APB) on February 14, 2003, which expanded the modernization requirements for the new GPS satellites to include an additional requirement for "flex-

---

[15] Hura et al., 2011.

[16] Hura et al., 2011.

[17] GAO, 2009b. GPS IIA and IIR satellites have one civil signal, C/A code on L1 frequency band, and two military signals, P(Y) code on L1 and L2 frequency bands.

[18] Hura et al., 2011.

ible power."[19] This was followed by another contract modification awarded in September 2003 for including both the same modernization and the flexible power requirement.

The cost growth history for the GPS IIF program is shown in Figure 2.3 and is based on the PM's EAC reported in the SARs since the original contract. Because no significant cost growth occurred before the introduction of the modernization requirements, the cost data between the date of the original contract and 2002 are not shown. There was no SAR submission for December 2008. The modernization accounted for a significant portion of the ensuing cost growth by increasing scope and contributing to technical issues. First, by September 2002, the baseline was recalculated, and the expected cost rose by $240 million (FY 2013 dollars) for the six satellites—a 43 percent cost growth. The requirements for flexible power were the largest contributor to the cost growth of $250 million in 2003. Additionally, GPS IIF encountered many difficulties when implementing the requirement changes in the midstream of the program. For example, the solar panel size was increased and its design was altered to incorporate more advanced gallium arsenide cells to provide increased power. GPS IIF also struggled with parts quality issues that led to test failures, adding more costs. These

Figure 2.3
History of GPS IIF Cost Growth

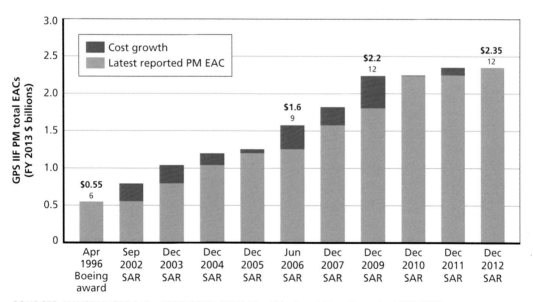

SOURCES: NAVSTAR GPS SARs, 1996–2012; SBIRS Monthly Acquisition Reports, 1996–2012.
NOTE: Numbers below dollar amounts refer to cumulative number of satellites ordered.
RAND MG1171/7-2.3

---

[19]   Hura et al., 2011. "Flexible power" involved swapping power between legacy and the new military M-code signal as needed in a jamming environment to leverage the signal strength of one signal or the other (NAVSTAR GPS SAR, December 2002).

technical issues added to the cost increases and, by December 2005, the cost of the six satellites had more than doubled.

In 2006, the Air Force decided to reduce the GPS IIF buy to 12 satellites rather than the originally planned 33 satellites. This reduction in quantity increased the unit costs of the satellites.[20] The Air Force exercised the option and placed satellites SV-7 through SV-9 on contract in 2006. The program continued to have technical issues with the navigation payload and parts quality issues, requiring redesign, rework, and additional testing, raising the EAC to $1.8 billion in 2007. By the end of 2009, the EAC had grown to $2.25 billion as a result of adding satellites SV-10 through SV-12 and continued technical difficulties associated with environmental testing as well as resolving interface issues between the navigation payload and nuclear detonation detection system. In the end, the weight of the redesigned satellite doubled to accommodate the new requirements.

The GPS IIF cost growth moderated afterward, with additional cost growth of $100 million in 2011, primarily to resolve multiple navigation payload parts and clock failure issues. One cesium clock (one of three atomic clocks onboard) on the second GPS IIF satellite experienced an on-orbit failure, and subsequent GPS IIF satellites have implemented improved rubidium clocks.[21]

Table 2.2 shows the composition of the sources of GPS IIF program total cost growth of $1.8 billion in terms of technical difficulties and the programmatic factors (namely, modernization requirements and quantity changes). The sources are listed in a decreasing order of level of contribution to cost growth (i.e., from the highest to the lowest cost-growth source). We found that the total cost growth attributed to technical issues is comparable to that attributed to programmatic changes.

Table 2.2
Sources of GPS IIF Program Total Cost Growth (in Decreasing Order of Percentage Contribution)

Technical Cost-Growth Source	Programmatic Cost-Growth Source
Payload and system-level environmental testing anomalies	Added modernization (civil, M-codes, and flex power)
Late navigation payload hardware deliveries and parts quality/rework issues	Exercised options for SV-7 through SV-12
Multiple navigation payload parts issues and clock failure	
Navigation payload and nuclear detonation interface compatibility issues	

SOURCES: NAVSTAR GPS SARs, 1996–2012; GPS IIF Monthly Acquisition Reports, 2005–2012.

---

[20]   GAO, 2007a.

[21]   Cooley, 2013.

The program schedule was adversely affected as a result of the introduction of additional requirements in midstream of the program. Figure 2.4 shows a history of the schedule for the first GPS IIF satellite based on the satellite delivery date. The solid bar corresponds to the planned delivery dates, and the cross-hatched bar refers to the actual delivery date. The schedule continued to slip after the addition of the modernization requirements and ultimately delayed the launch of the first GPS IIF by five years, from the originally planned date of June 2005 to March 2010. As of August, 2014, seven GPS IIF satellites were in orbit and the eighth satellite is slated for launch during the fourth quarter of 2014.

Advanced Extremely High Frequency Program

The advanced extremely high frequency program was initiated in 1999 as a follow-on to Milstar II to provide higher-capacity, survivable, jam-resistant, worldwide, secure communication capabilities for strategic and tactical warfighters.[22] The AEHF program appeared to be poised for success. Its incumbent team was experienced, most of its technologies were mature, and it had the advantage of applying lessons learned from

Figure 2.4
GPS IIF Program Schedule Overrun History

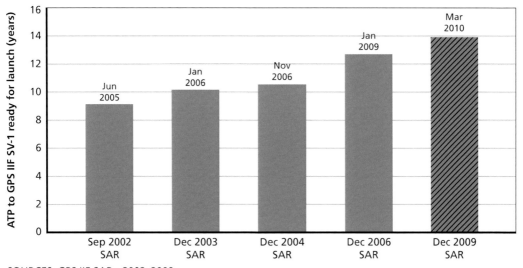

SOURCES: GPS IIF SARs, 2002–2009.
RAND MG1171/7-2.4

---

[22] In 1993, initial plans were laid to acquire the more advanced follow-on extremely high frequency program, Milstar-III/AEHF. In 1995, the Milstar-III/AEHF was decoupled from the Milstar program and became the AEHF program (Hura et al., 2007).

Milstar.[23] But the program suffered from several risk factors: aggressive schedule acceleration (a result of a Milstar launch failure in April 1999),[24] immature cryptographic equipment provided by the National Security Agency (NSA), and unexpected changes in military satellite communications (MILSATCOM) constellation requirements as a result of a cancellation of another MILSATCOM program.

Shortly after the 1999 system definition award, its acquisition strategy was changed to prevent a coverage gap in the wake of the aforementioned Milstar loss. A "National Team" composed of three contractors was put together in the hopes that it could build an AEHF satellite 18 months faster.[25] This attempted acceleration led to requirements that were not finalized and multiple engineering change proposals.[26] Further, the plan to take advantage of contractor commercial experience and to follow commercial practices turned out to be a high risk in that a schedule based on a commercial analog was inappropriate for a satellite as complex as the AEHF. For example, commercial communications satellites do not involve cryptographic units and nuclear hardening. The instability in the AEHF constellation size (a result of MILSATCOM architecture issues) also led to costly stop-start of the production line. Five AEHF satellites were originally called for in the protected MILSATCOM architecture, but only two AEHF satellites were initially placed on contract. Then, the Transformational Satellite Communications System (TSAT) program began in 2004, and the AEHF constellation size was reduced to three by March 2005, with the expectation that the TSAT satellites would be on orbit by FY 2009 and would complete the protected MILSATCOM constellation. The third AEHF satellite was not placed on contract until 2006. As the TSAT program stalled, a fourth AEHF satellite was planned for in 2008, although it was not added to contract until 2010. After the 2009 cancellation of TSAT because of concerns over costs and risks,[27] the AEHF constellation size changed from four satellites to six, and the contracts for the fifth and sixth satellites were not awarded until 2013.[28]

Figure 2.5 shows the history of cost growth incurred in the AEHF program in terms of the PM EAC since the system development and demonstration (SDD) award. There was a minimal change in the EAC in the December 2002 SAR, and thus it is not shown in the figure. There was no SAR submission for December 2008, but a midyear reporting was required in 2008 because of a Nunn-McCurdy breach. Early

---

[23] Hura et al., 2007.

[24] Titan IVB failed to boost the third Milstar into proper orbit (Whitley, 1999).

[25] Gansler, 2000.

[26] Hura et al., 2007.

[27] Brinton, 2009.

[28] In December 2012, the Air Force awarded a fixed price contract to Lockheed Martin, giving it $1.94 billion for AEHF-5 and AEHF-6. The cost growth that resulted from this scope increase is not reflected in the December 2012 SAR, because the contract was not definitized until October 31, 2013 (AEHF SAR, 2013).

in the program, many technical issues contributed to significant cost growth. These included technical problems with the NSA cryptographic units, switching from field-programmable gate arrays (FPGAs) to application-specific integrated circuits (ASICs), reworking super-high-frequency array panel units, and correcting digital processor power converter designs. Efforts to fix these technical problems led to the first AEHF program Nunn-McCurdy breach of significant cost growth in 2004. After the program was rebaselined, the new EAC grew to $4.7 billion as reported in the December 2005 SAR.

In 2006, the program EAC grew to about $5.4 billion when a third satellite and tasks on the ground segment, launch operations, and sustainment were added. Two years later, the program reported a second Nunn-McCurdy breach when the cost grew to about $7.1 billion.[29] During this period, a significant portion of the cost growth was attributed to technical difficulties. Many failures emerged during the integration and testing of SV-1; in several cases, they involved deeply buried components that were difficult to remove and correct. Because full military standards were not required for all parts, the level of component testing and assurance was inadequate. The ground software development also experienced difficulties. Many deficiencies were found in the Milstar backward-compatibility software, and resolving these issues also contributed to the cost growth in 2008. Further, the program office was planning for a fourth AEHF, because the

Figure 2.5
AEHF Program Cost Growth History

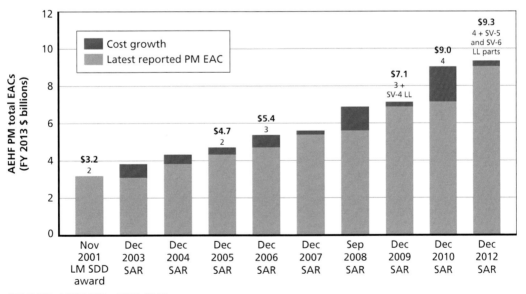

SOURCES: AEHF SARs, 2001–2012.
NOTE: Numbers below dollar amounts refer to cumulative number of satellites ordered.
RAND MG1171/7-2.5

---

[29] The September 2008 SAR was in response to the second Nunn-McCurdy breach.

TSAT program was stalling. The additional cost of resolving design problems, reworking mission critical elements, and conducting additional testing of SV-1 and SV-2 combined with the cost of SV-3 and the long-lead items for SV-4 to plan for procurement of SV-4 led to the second Nunn-McCurdy breach of critical cost growth.

Since the second Nunn-McCurdy breach, the cost growth resulting from technical difficulties has been relatively low, possibly because the development work was coming to completion and the program was transitioning into production. In 2010, when the first AEHF satellite was launched, it experienced a failure in the satellite's liquid apogee engine that was to maneuver the satellite toward its operational orbit.[30] Resolving this anomaly contributed to further cost growth. Most of the cost growth since 2009 has resulted from buying more satellites. The fourth AEHF satellite was added in 2010, and the long-lead items for the fifth and sixth satellites were added in 2012.

The composition of the technical and programmatic sources of total cost growth of $6.1 billion is shown in Table 2.3 in decreasing order of contribution (i.e., from the highest to the lowest cost-growth source). Similar to the SBIRS program, cost growths attributed to technical difficulties and programmatic factors are split about evenly. The technical difficulties were primarily associated with additional testing, redesign, and rework of the first two satellites. The cost growth resulting from programmatic factors is primarily associated with adding four more satellites (SV-3 through SV-6) since the original contract.

The history of schedule delays for the AEHF program is shown in Figure 2.6 in terms of the time line to AEHF's initial operational capability (IOC). The IOC for AEHF entailed two AEHF satellites operating at extended data rates (XDRs) with backward-compatibility to legacy Milstar systems. The figure shows that the IOC

Table 2.3
Sources of AEHF Program Total Cost Growth (in Decreasing Order of Percentage Contribution)

Technical Cost-Growth Source	Programmatic Cost-Growth Source
Additional system rework and testing	Added SV-3 and SV-4 plus long-lead parts for SV-5 and SV-6
Spacecraft bus subsystem parts qualification (e.g., switching from FPGAs to ASICs)	Delays resulting from government furnished equipment NSA cryptographic units
Milstar backward-compatibility software issues	Additional ground segment, launch, and sustainment tasks
Payload assembly parts qualification and retesting	Additional radiation hardening and Milstar backward-compatibility
SV-1 apogee engine anomaly resolution	

SOURCES: AEHF SARs, 2001–2012; GPS IIF Monthly Acquisition Reports, 2005–2012.

---

[30] "AEHF-1 Arrives at Its Operational Orbit 14-Months Journey," 2011.

has continued to slip since the beginning of the program, with the latest estimate of planned IOC of June 2015. The first AEHF satellite reached its intended orbit in October 2011 (following 14 months of maneuvering the satellite after the spacecraft's liquid apogee engine failure).[31] The second AEHF satellite was launched in May 2012 and was turned over for operations in November 2012, and the third AEHF satellite was launched on September 18, 2013. However, software issues in the AEHF mission control system delayed the XDR capability to June 2015.

Wideband Global SATCOM[32]

The WGS was originally undertaken to provide unprotected—basically commercial-grade—communications to U.S. forces to fill the gap between the Defense Satellite Communication System (DSCS) III and more powerful satellites such as TSAT. WGS is a commercial Ka-band satellite (based on the Boeing/Hughes 702 bus) plus a few cryptographic modules and frequency cross-banding to work with existing DoD X-band and Ka-band user terminals.[33] Thus, such satellites were expected to

Figure 2.6
AEHF Program Schedule Delay History

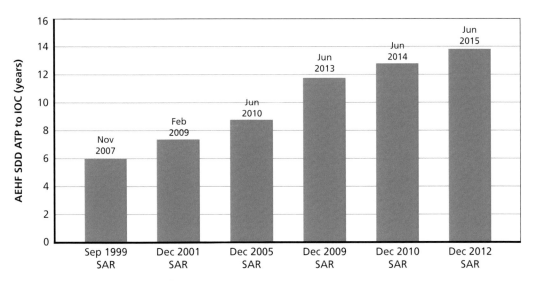

SOURCES: AEHF SARs, 1999–2012.
RAND MG1171/7-2.6

---

[31] "AEHF-1 Arrives at Its Operational Orbit 14-Months Journey," 2011.

[32] The material for this section was extracted from Blickstein et al., 2011, notably Chapter Six.

[33] Each of the first three WGS satellites is equipped with ten Ka-band antennas. Eight of them provide narrow coverage, and the other two provide wider area coverage. The coverage of the oval-shaped beam coming from the Ka-band antennas spans 600 miles, and the beam can be steered through a gimbal system, a feature not available with previous-generation satellites.

need only minimal research and development and to be priced similar to their commercial counterparts. However, the unit cost of the WGS satellites increased significantly since the first block of WGS satellites was placed on contract, as we will discuss below.

Figure 2.7 displays a summary of the WGS program increases in total contractor EACs from the initial Boeing firm fixed price (FFP) contract for Block I satellites beginning in January 2001 through the December 2012 SAR in FY 2013 dollars.[34] There was a minimal change in the EAC in December 2003 SAR and thus it is not shown in the figure, and there was no SAR submission for December 2008.

The scope of the initial FFP contract awarded to Boeing in January 2001 included nonrecurring design effort and initial advance parts in support of the first three satellites. Over the following two years, three satellites were added. In the 2004 and 2005 time frame, the contractor experienced technical difficulties that

Figure 2.7
History of the WGS Program Cost Growth

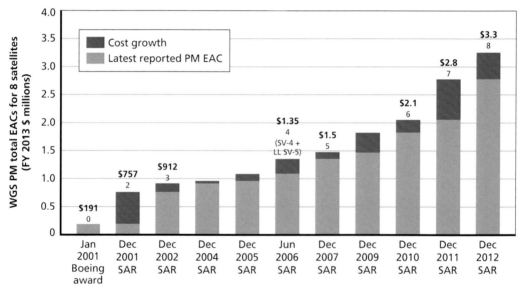

SOURCES: WGS SARs, 2001–2012.
NOTE: Numbers below dollar amounts refer to cumulative number of satellites ordered.
RAND MG1171/7-2.7

---

[34] The figure and the total EAC increases do not include any cost increases in the production and delivery of WGS Block II SV-6, which is being procured separately and paid for by Australia in exchange for accessing a portion of the WGS constellation bandwidth. The WGS total EAC through December 2012 also does not include any cost increases for procuring the Block II follow-on SV-9 satellite, which is being procured and funded separately under a cooperative agreement through an international partnership and memorandum of understanding (MOU) with Canada, Denmark, Luxembourg, the Netherlands, New Zealand, and the United States. The MOU was signed for the procurement of SV-9 in exchange for access to the WGS constellation and was put under contract to Boeing in January 2012.

contributed to the cost growth. First, the design and manufacturing of the phased array antenna turned out to be more difficult than anticipated. It required scaling the commercial Ka-band technology to the X-band (8–12 GHz) to extend military services (i.e., those provided by DSCS). Boeing also added extra solar panels to the original design, which added weight to the satellite, causing changes in the launch vehicle. The contractor had also used incorrectly sized fasteners on the spacecraft, and thus new fasteners had to be installed.[35] Some of the cost growth is attributed to adding a new requirement. The program added a radio frequency bypass capability to allow transmission of airborne intelligence, surveillance, and reconnaissance unmanned aerial vehicle (UAV) mission data at a much greater data rate.

In 2006, the Air Force ordered a second block (WGS Block II) of three satellites (including one satellite [SV-6] essentially purchased by Australia) when DoD found itself needing more satellites as the TSAT program stumbled and users found ways to fill the bandwidth they received and asked for more. However, the price of WGS satellites rose dramatically, from $240 million to $380 million each (in FY 2001 dollars). The Air Force argued that Boeing's first price was unreasonably low, particularly in light of later weaknesses in the market for commercial communication satellites; $380 million was claimed to be the true cost.The EAC grew to $1.5 billion by the end of 2007 as a result of the additional two satellites, parts obsolescence, and WGS-1 launch vehicle integration issues. In 2009, the EAC grew by another $300 million. During this time frame, the scope was increased to add launch service and launch site processing facilities tasks. Contractor difficulties associated with technical and supplier issues also contributed to the cost growth. These issues included late delivery of parts, transponder anomalies, flight software development issues, delamination of solar array panels, and spacecraft control processor unit anomaly during thermal vacuum testing. Resolving some of these issues continued through 2011. In 2010, because WGS services remained popular, DoD planned a buy of six more satellites (commitments for actual purchase would be made on a year-to-year basis), referred to as a WGS Block II follow-on. But the unit price of the satellites came in at $580 million, a second substantial increase. [36] By 2012, three additional WGS satellites and a requirement for a digital channelizer were added, and the EAC grew to $3.3 billion. The cost growth from increases in the satellite unit price was the dominant cost-growth source during this time frame.

Why did the unit satellite price continue to increase when satellites 7 through 12 were essentially the same as the first six? Multiple factors contributed to this cost growth, many having to do with accounting issues and the high inflation rates of the

---

[35] U.S. Department of Defense, "Selected Acquisition Reports: Wideband Gap Filler System," 2005.

[36] Blickstein et al., 2011, provides details of the RAND WGS root-cause analysis for the OUSD (AT&L) Performance Assessments and Root Cause Analysis (PARCA) office. The report attributes around $73 million in the WGS Block II follow-on SV-7 and SV-8 unit cost per satellite increase (in FY 2011 constant dollars) over Block II SV-4 through -6 satellites' comparable unit cost to production gap–related causes, including having less long-lead parts storage and additional factory restart follow-on satellite costs.

satellite industry as a whole. Appendix B provides further discussions on the WGS example and the overall cost effects and issues associated with differences between defense space systems inflation rates and standard Office of the Secretary of Defense (OSD) escalation indices. But two parts to the answer merit note. First, over time, Boeing shifted its commercial satellite offerings from its 702HP (high-power) bus to its 702MP (medium-power) bus. As WGS specifications drifted further from commercial specifications, it was less able to take advantage of the commercial industrial base; WGS became an increasingly military satellite. This shift from commercial to military was further underlined by the tendency of component prices to escalate, because the alternative was for specialized defense contractors to leave the business because of limited demand for the specialized components. Second was the 2½-year production gap between the second and third tranches arising from varying requirements for the constellation's size,[37] and cancellation of TSAT exacerbated matters.

Table 2.4 lists the technical and programmatic cost-growth sources that contributed to the total program cost growth of $2.3 billion in a decreasing order of contribution. We found that purchasing additional WGS satellites was the dominant contributor to total program cost growth. The cost growth resulting from technical issues was moderate.

The schedule history of the first WGS satellite is shown in Figure 2.8. WGS-1 was launched in October 2007 and was declared IOC in January 2009, four years later than originally planned. The remaining two satellites in the first block of WGS were launched in 2009.

Global Positioning System III

GPS III is the next generation of GPS satellites to complete the modernization of the GPS constellation. GPS III will sustain the GPS constellation by replacing aging GPS satellites with satellites that have improved accuracy, integrity, and assured availability for both civilian and military users worldwide. The original GPS III acquisition strategy was to provide these capability enhancements in three increments shown below (IIIA, IIIB, and IIIC):[38]

- Eight GPS IIIA satellites would provide an internationally compatible new civil signal (L1C) and an increased M-code anti-jam power with full earth coverage.
- Eight GPS IIIB satellites would include near-real-time command and control crosslinks.

---

[37] Blickstein et al., 2011.

[38] U.S. Department of Defense, "Selected Acquisition Reports: GPS-IIIA," June 2008; GAO, 2009a.

Table 2.4
Sources of WGS Program Total Cost Growth (in Decreasing Order of Percentage Contribution)

Technical Cost-Growth Source	Programmatic Cost-Growth Source
Parts design and manufacturing difficulties (e.g., phased array)	Added Block III and digital channelizer requirement
WGS-1 launch vehicle integration issues	Added Block II (SV-4 and SV-5)
Anomalies during thermal vacuum testing	Added launch service and site processing facilities tasks
Transponder anomalies	Switch in launch vehicle
Block II flight software development issues	Added airborne intelligence, surveillance, and reconnaissance UAV bypass requirement
Parts obsolescence issues	
Miscellaneous technical issues (e.g., solar array panel delamination)	
Delays in parts	

SOURCES: WGS SARs, 2001–2012; WGS Monthly Acquisition Reports, 2009–2012.

Figure 2.8
WGS Program Schedule Delay History

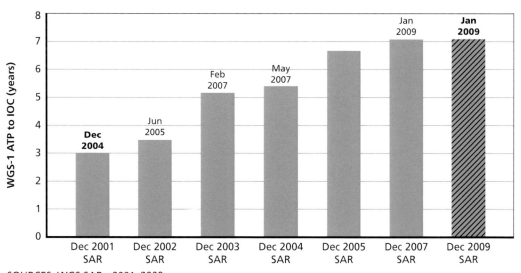

SOURCES: WGS SARs, 2001–2009.
RAND MG1171/7-2.8

- 16 GPS IIIC satellites would include a high-powered spot beam that increased anti-jamming capability.

In 2008, Lockheed Martin was awarded the $1.4 billion GPS IIIA contract to produce a prototype, the GPS III Non-Flight Satellite Testbed (GNST), and two GPS III satellites, with an option for ten more.[39] To prevent the types of engineering issues discovered on other programs late in the manufacturing process or even on orbit, the Air Force implemented a "back-to-basics" acquisition approach for GPS III. This approach started circa 2006 as DoD space programs were experiencing cost growth and delays, and the difficulties of the 1990s' acquisition reform initiatives became widely recognized. As described in its implementation memo, "Back to Basics and Implementing a Block Approach for Space Acquisition,"[40] this approach advocated the following:[41]

- the use of a disciplined acquisition approach and an experienced, high-quality, technically educated, government workforce actively engaged in all aspects of the enterprise
- clear and achievable requirements
- disciplined systems engineering
- effective management
- appropriate resources
- stabilized and aligned requirements and resources
- risk assessments and risk mitigation
- program cost estimates at the 80 percent confidence level by Milestone B
- a block approach based on incremental deliveries providing more rapid initial capability based on proven technologies, with later blocks supported by investing in science and technology development
- coordination with the user as to the timing of blocks and the capabilities to be delivered.

For the Air Force, "back to basics" meant reintroducing such earlier practices as government oversight, qualified personnel, and stable program funding,[42] coupled

---

[39] The GNST is a fully functional prototype satellite with the navigation payload and fully functional nonflight boxes used as a pathfinder for design, integration, and tests.

[40] Sega, 2007.

[41] Going back as far as 2004, Air Force senior leaders have been advocating the back-to-basics approach. TSAT was to pioneer the back-to-basics acquisition approach (Sirak, 2006).

[42] Space and Missile Systems Center, 2004.

with serious system engineering and testing using time-tested knowledge that incorporated lessons learned.[43]

Even with the back-to-basic approach, the GPS III program began with some schedule risk. The program began with a six-year schedule between the start of engineering development (February 2008) and first spacecraft delivery (May 2014)—more aggressive than prior GPS program schedules. This accelerated schedule was necessary because failure to meet it introduced the possibility of gaps in GPS capability as aging satellites reached their lifetime operating limit—a fact exacerbated by the six-month delay in the start of the GPS III program.[44] In addition, Lockheed Martin had to assemble a workforce to implement the program from the ground up, since it was not the builder of the GPS Block IIF satellites.[45]

Figure 2.9 shows the history of cost growth incurred in the GPS III program in terms of the PM EAC since the original contract award. There was no SAR submission for December 2008. Overall, GPS III's cost growth appears moderate compared with that of SBIRS, AEHF, and GPS IIF at comparable points in their history. Similarly, the schedule growth of the first GPS III satellite was about five months when the December 2012 SAR was released.

In 2009 and 2010, GPS III experienced cost growth associated with meeting navigation payload parts reliability and performance requirements. In 2011, the total program EAC decreased, primarily because an adjustment was made to correct an inaccuracy in the initial contract price. However, there was a cost-growth component associated with technical issues. Some FPGAs in the satellite bus's telemetry, tracking, and command (TT&C) subsystem did not meet space-qualification tests, and resolving this issue increased cost. Radiation lot acceptance testing also proved unexpectedly difficult. In January 2012, the Air Force ordered the second set of two satellites for an additional $238 million.[46] Adding launch and checkout capability tasks further increased the government scope and thus cost. The cost growth attributed to contractor difficulties in 2012 included satellite bus system hardware issues (namely, the inertial measurement unit and scalable power regulator unit) and software issues with the mission data unit. The SAR also reported that the cost growth could be partly attributed to decreased economies and efficiencies resulting from a reduced production rate from four SVs per year to two SVs per year.[47]

---

[43] In addition, in the latter half of 2007, SMC created its Program Management Assistance Group, whose function was to assure thoroughness and consistency of program baseline documents, including systems engineering and test documents.

[44] GPS III SAR, June 2008.

[45] GAO, 2009a.

[46] Lockheed Martin, undated.

[47] December 2012 SAR.

Figure 2.9
GPS III Cost Growth History

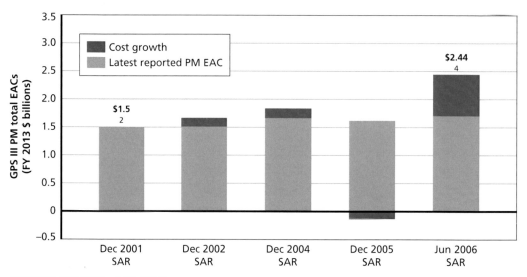

SOURCES: WGS SARs, 2001–2012.
NOTE: Numbers below dollar amounts refer to cumulative number of satellites ordered.
RAND MG1171/7-2.9

The GPS III acquisition strategy is being modified to streamline Blocks IIIB and IIIC into one increment called SV 9+.[48] A reduction in requirements for later increments to lower the cost and risk of the program and broader options for the follow-on production satellites are under consideration.[49] As of June 2014, the Air Force is looking for alternative sources to compete with Lockheed Martin for the continuing production of up to 22 satellites.[50]

Table 2.5 lists the technical and programmatic cost-growth sources that contributed to the total program cost growth of $0.94 billion in a decreasing order of contribution. About two-thirds of the total cost growth is attributed to technical difficulties, with the largest source of cost growth being navigation payload issues.

Summary

Table 2.6 summarizes the performance of the five selected DoD space programs. Note that SBIRS, GPS IIF, AEHF, and WGS were initiated in the 1990s' acquisition reform

---

[48] GAO, 2013c.

[49] GAO, 2013c; staff discussions with personnel in the Office of the Under Secretary of the Air Force for Acquisition about space acquisition and the status of the GPS III program, 2013.

[50] Peck, 2014.

Table 2.5
Sources of GPS Program Total Cost Growth (in Decreasing Order of Percentage Contribution)

Technical Cost-Growth Source	Programmatic Cost-Growth Source
Navigation payload parts issues in meeting reliability specifications and performance requirements	Exercised SV-3 and SV-4 production options and procured long-lead items
Bus parts technical issues (e.g., inertial measurement unit, power regulator unit rework)	Added launch and checkout capability tasks
Mission data unit software development issues	
Bus TT&C subsystem FPGA difficulty in meeting design requirements	

SOURCES: GPS III SARs, 2008–2012; GPS III Monthly Acquisition Reports, 2008–2012.

era, and GPS III was initiated in the "back-to-basics" era.[51] The first four programs experienced major cost growth and schedule overruns, extended cycle times, and instability in quantity. By contrast, cost and schedule growth for GPS III appears to have been more moderate than the other programs, so far. SBIRS and AEHF incurred multiple Nunn-McCurdy breaches; WGS incurred one such breach. The GPS IIF program would have breached but for technicalities.[52]

SBIRS-GEO, GPS IIF, AEHF, and WGS all experienced delays in the launch of their first satellites. Although six years from ATP to first launch of WGS Block I satellite seems relatively short compared with the cycle times of SBIRS, GPS IIF, and AEHF, WGS Block I was a commercial-like buy, which should not have taken more than three years, a typical production cycle for a commercial satellite.

Fortunately, the satellites that such programs produced do meet the performance requirements. The few technical problems that occurred with the satellites on-orbit did not jeopardize the space mission that the satellites supported. Nevertheless, the full capability status for SBIRS, GPS IIF, AEHF, and WGS requires that the constellations be fully populated and that their ground control segment be fully operational. GPS III, the other program we studied, has fared much better so far, perhaps in part because of its back-to-basics acquisition approach.

---

[51] The back-to-basics era began circa 2006. See Chapter Five for more information on back-to-basics.

[52] GPS IIF did not incur a Nunn-McCurdy cost breach, because the PAUC and the APUC estimates for GPS IIF have not been separately reported in the NAVSTAR GPS program SAR since the beginning of the GPS IIF ATP through the latest December 2012 SAR. For SAR program budget cost-reporting purposes, GPS IIF is considered as part of the NAVSTAR GPS enterprise program, which includes 13 GPS IIR, 8 GPS IIR-M, and 12 GPS IIF satellites and the ground mission operational control segment (OCS). As a result, the PAUC and the APUC cost growth of the NAVSTAR GPS program is based on a total budget estimated for acquiring 33 satellites and not just the 12 GPS IIF satellites.

Table 2.6
Performance of Selected DoD Space Programs

Program (Space Mission)	Milestone B (to 1st Launch)	Normalized Cost Growth (%)[a]	Schedule Growth (%)[b]	Change in Quantities	Satellite Performance Shortfalls
SBIRS (missile warning)	1996 (15 years to GEO-1 launch)	376	157 8.8 year GEO-1 launch delay	5 GEO-2 HEO to 4 GEO-4 HEO + GEO-5 and -6 initial NRE + long-lead parts	Onboard communications issue on GEO-1
GPS IIF (PNT)	1996 (14 years)	237	52 4.75 year SV-1 launch delay	6–12 satellites	Clock failure on 2nd GPS IIF
AEHF (protected MILSATCOM)	2001 (9 years)	105	93 6.5 years IOC delay	2–4 satellites + SV-5 and 6 long- lead parts	Apogee engine failure on AEHF-1
WGS (wideband MILSATCOM)	2001[c] (6 years)	142	136 4.1 year IOC delay	3–8 DoD satellites	None
GPS III (PNT)	2008 (N/A)	45	7 5 month SV-1 delay	2–4 satellites	Not yet launched

SOURCES: Program SARs from Milestone B to 2012; GAO, 2012.

[a] Normalized cost growth is the percentage increase in the PM's contractor total EACs from ATP through the December 2012 SAR over the initial Milestone B contract price (both in FY 2013 dollars), adjusted to account for the quantity change (i.e., cost growth resulting from quantity increase was excluded).

[b] Except for AEHF and WGS delays in meeting expected program IOC, schedule program growth is the percentage increase from ATP to the first actual (or latest reported) satellite launch date over the original planned date.

[c] Denotes the year of initial WGS contract award date.

The four programs—SBIRS, AEHF, GPS IIF, and WGS—experienced major cost growth and schedule delays arising from difficulties in technology development, integration issues, parts quality issues, and obsolescence that led to costly redesign and rework. These programs took on the following risks that contributed to these difficulties and inefficiencies including those listed below:

- introducing immature technologies
- accelerating program schedules
- changing requirements midstream
- inadequate testing and systems engineering
- changing buy quantity and inefficient buying practice, causing long production gaps resulting in obsolescence.

Chapters Three and Four examine systemic issues and key driving factors that led the programs to take the high-risk acquisition approach.

CHAPTER THREE

Effects of the 1990s' Space Acquisition Environmental Factors on the Programs

Fluid national security issues, budgetary constraints, and the health and performance of its industrial base shape the space acquisition environment. The end of the Cold War sharply changed U.S. national security strategy and hence the defense budget in the 1990s—the decade when many current space programs began. The defense budget shrank and preparing for regional conflict received increased emphasis.

In this chapter, we look at how these changes affected the management and execution of the major DoD space programs (except for GPS III, which started in a later era). Specifically, we identify how shifting and growing requirements, acquisition reform initiatives, and speculations about the commercial space market contributed to a high-risk approach to space acquisition.

Shift in Requirements Increased Complexity[1]

During the Cold War, the United States was faced with a nuclear superpower equipped not only with weapons of mass destruction but also with multiple means of delivering those weapons to targets throughout the world, including homeland America. Thus, Mutual Assured Destruction became a key element of U.S. strategic defense strategy. Soviet Union intercontinental ballistic missiles (ICBMs), submarines, and long-range bombers, armed with nuclear warheads and weapons, respectively, were predominant threats to the U.S. homeland. Space programs and their mission capabilities were essential in providing persistent and flexible access to denied areas that could not be obtained by other means. Satellites were designed to detect and monitor the developments, activities, and postures of Soviet ballistic missiles, long-range bombers, submarines, and nuclear weapons. They led to the development of space-based imaging systems such as Corona (1958) and, in 1962, the Defense Meteorological Satellite Program—the latter to enable more efficient imagery collection (e.g., to avoid sending spy planes over the Soviet Union on cloudy days).

---

[1] The material for this section was extracted from Hura et al., 2007, notably Chapter Two.

The Defense Support Program was started in the mid-1960s to generate early warning alerts of Soviet missile launches, allowing U.S. retaliatory forces (strategic bombers, ICBMs, and submarine-launched ballistic missiles) to be launched before being destroyed. Prosecuting the Cold War also required the ability to relay secure strategic communications from the National Command Authority throughout the world and with assured delivery (e.g., Emergency Action Messages). Military satellite communications also were developed to provide wide-bandwidth conduits for increasing amounts of data and information supporting users worldwide. Tactical operators, nevertheless, found useful services from national space systems built to support strategic requirements.

The end of the Cold War led to a greater emphasis on regional conflicts, which demanded more space support to tactical operations. Furthermore, the value of space capabilities in tactical operations was recognized in the first Gulf War. The demand for satellite support to tactical operations shifted the emphasis in designing satellite systems from strategic to tactical uses.[2] Because the strategic missions were never abandoned, the addition of new technical requirements only increased the number of stakeholders as well as space system users. Not surprisingly the decision chain became more complex and longer as the increase in the number and diversity of stakeholders increased the likelihood of conflicts of interest, such as conflicts in priorities, which can be difficult and time-consuming to resolve. GAO has also frequently stressed that the disjointed leadership structure helps incubate current problems because of low accountability and visibility, and it also inhibits conflict resolution between stakeholders regarding requirements and funding.[3]

Also unsurprising, the list of requirements also grew, as a 2003 DSB report observed. This growth, in turn, led to "reduced program manager flexibility because of the increased number of key performance parameters (KPPs) needed to satisfy and maintain the support of the expanded constituency."[4] More requirements meant more complex systems integration with multiple mission requirements placed on single satellites. A demand for increased performance to meet the additional tactical requirements also contributed to the design of more complex satellites. These combined requirements eroded the original assumption that future DoD space programs would be just upgrades or modernizations rather than innovations. There was a fundamental need to integrate new state-of-the-art, immature technologies to implement the complex and demanding requirements generated in this time period.

Table 3.1 summarizes how tactical requirements were added to various satellite programs. SBIRS, nominally just a DSP follow-on, has four primary missions rather than DSP's single mission (strategic missile warning); it supports both strategic and

---

[2]   Pawlikowski, Loverro, and Cristler, 2012.

[3]   GAO, 2013b.

[4]   OUSD (AT&L), 2003.

Table 3.1
Increased Tactical Mission Requirements

Space Program	Additional Requirements Compared to Predecessor Systems
SBIRS	Four primary missions compared to one primary mission on DSP: Missile warning (strategic and tactical), missile defense (strategic and tactical), technical intelligence, and battlespace characterization
GPS IIF (post-modernization)	Increased jam resistance New military code, M-code, for military jam resistance Flexible power for military signals
AEHF	Increased tactical protected MILSATCOM capability XDR package data throughput (8 Mbps) compared to Milstar II (1.5 Mbps) 400 Mbps capacity (50 channels of XDR) compared to Milstar II's 40 Mbps (32 channels of medium data rate)
WGS	Increased tactical MILSATCOM capability Communications in tactical X-band and Ka-band compared to X-band only on DSCS III Increased capacity (2.1–3.6 Gbps) compared to DSCS III (250 Mbps) X-Ka crossbanding
GPS III	Increased jam resistance Increased M-code anti-jam power with full earth coverage

theater missile warning and missile defense. Its 19 KPPs range from DSP-like missile warning (with eight KPPs) to missile defense (five KPPs) to technical intelligence (five KPPs) to battle space characterization (one KPP). DSP satellites carried only scanning sensors; SBIRS-High would add staring sensors. This complexity in requirements contributed to poor understanding of the flow-down of KPP requirements, technology development, and integration complexity, as discussed in Chapter Two.[5]

As a result of the GPS modernization, GPS IIF satellites had a new set of KPPs that were introduced midstream in the program. These requirements would improve the military use of GPS for tactical operations (in addition to other improvements for civil users): (1) a new military code, M-code, that was more jam resistant, and (2) flexible power for military signals, which enables increasing signal strength as needed in a jamming environment.[6] These requirements meant new technologies.

Although the GPS III program began in a later era, the shift in the emphasis toward tactical operations called for an improved anti-jam capability to protect its military code.

The goal of AEHF was to build a smaller, lighter, and less-expensive satellite than its predecessor, Milstar, but with added capability, notably support to the tactical user.[7] To meet the tactical protected MILSATCOM requirements, the AEHF payload added an XDR waveform to those already carried on Milstar-II—for a

---

[5] This is one finding from an Independent Review Team in April 2002 reported in the June 2002 SAR.

[6] Hura, 2011.

[7] Pawlikowski, 2006.

low data rate and a medium data rate. XDR allowed much higher throughput for tactical military communications, such as real-time video, battlefield maps, and targeting data.[8] The XDR payload provides communications data capacity that is ten times more, made possible by more channels, each of which has data rate that is about five times higher than the medium date rate payload on Milstar II.

As for WGS, circa 1995, DoD projected that its needs for high-capacity satellite communications would grow, driven by its greater dependence on reach-back capabilities for deployed forces.[9] Insofar as the DSCS replacement would not be ready until 2006, DoD had to rely more on leased commercial satellite communications (SATCOM) capability until then. Hoping to leverage the commercial (SATCOM) boom of the late 1990s, DoD decided to buy something similar for itself. Thus, the WGS program began as a gap filler.

As a nearly commercial off-the-shelf product, WGS was supposed to buy the most capability for the dollar; it was not designed using traditional bottom-up analyses of warfighter needs.[10] However, it was supposed to provide an X-band (primarily a military spectrum) capability to alleviate saturation problems experienced with the DSCS constellation.[11] Hence, WGS combined the X-band and the Ka-band capabilities onto a single satellite, leading to technical difficulties: Developing the X-band phased array antenna and the cross-banding between the X-band and Ka-band to enable communications between X-band users and Ka-band users turned out to be complex and more difficult than anticipated.[12]

Acquisition Reform

The acquisition reform of the 1990s played a significant role in the SBIRS, AEHF, and GPS IIF programs. In January 1993, the Clinton administration declared acquisition reform to be a major priority.[13] Corresponding DoD policy and regulatory changes created an acquisition environment with a primary focus on cost, but it dropped many key government responsibilities in acquisition in favor of industry self-regulation for space programs. These changes called for reduced regulatory oversight on defense con-

---

8   Northrop Grumman, 2013b.

9   GAO, 1997.

10   Hura, 2007.

11   Director, Operational Test and Evaluation, undated (a).

12   Although phased arrays had been built for commercial SATCOM at Ka-band frequencies, the WGS called for scaling the commercial Ka-band technology to the X-band to extend military services. This scaling proved to be more difficult than originally thought (Hura et al., 2007).

13   Ingols et al., 1998.

tractors and increased reliance on commercial products and commercial best practices for defense programs.[14]

The new acquisition guidelines emphasized the following:

- streamlining requirements and reducing boiler-plate specifications
- shifting from oversight of contractor activities to building government insight
- assigning total system performance responsibility to the contractor
- maximizing use of commercial processes, technologies, and products.

To streamline requirements, a June 1994 memo from the Secretary of Defense entitled "Specifications and Standards—A New Way of Doing Business," directed that DoD replace military standards and specifications with commercial standards and specifications.[15] The replacement of government "oversight" with "insight" prevented the government from reviewing and demanding changes to the details of a design (oversight), since design was seen as the responsibility and expertise of the contractor. This insight did not include the detailed reporting requirements for cost and schedule data that would enable sound cost estimates or permit assessing a contractor's performance. Combined, these initiatives were called Total System Performance Responsibility (TSPR); it became a guiding principle in government-contractor responsibilities. It was based on the assumption that "in theory, the more responsibility the government can turn over to a contractor under a TSPR strategy, the greater the potential benefits."[16] The principles behind the acquisition reform were not necessarily the problem, but overaggressive implementation of TSPR, as we will argue, introduced imprudent risks to the space programs.[17]

TSPR led to prioritizing affordability; it gave contractors control over the system specification (the technical requirements)—the prime input for the systems engineering process—and allowed them to manage their own costs with little or no oversight using commercial parts and processes. The result was reduced rigor in systems engineering, integration, test, and program management. To cut costs and schedule, contractors reduced engineering, testing, and oversight of their subcontractors and vendors.[18] Implementing these commercial practices led to a decreased emphasis on

---

[14] For more details on the acquisition reform, see Younossi et al., 2008, and Hura et al., 2007.

[15] Perry, 1994.

[16] White, 2001.

[17] A RAND report found that the Air Force was probably the strongest service proponent of acquisition reform, and the Air Force Acquisition Executive saw to it that the reform was applied on many programs—in particular, the SBIRS (Hura et al., 2007).

[18] In a 2003 article, then-Commander of SMC stated that TSPR also led to confusion between the prime contractor and its subcontractors as to "who was responsible for what."

established supplier management practices and procedures for the DoD space program, such as the following:[19]

- strict supplier management, such that prime contractors worked with and oversaw the systems engineering and test practices of their suppliers
- incoming inspections where subcontracted parts and subsystems were thoroughly examined and tested before being integrated into prime contractor products
- use of pedigree parts to allow each part to be identifiable and traced back to the production lot that produced it, enabling a faulty part to expose its production lot, which could then be examined to see if it contained other faulty parts or engineering or manufacturing processes.

Weakened systems engineering introduced significant risks, because the programs that started during this era involved integrating state-of-the-art but immature technologies. Integrating immature technologies can be difficult even with strong systems engineering, because such technologies may change greatly as they mature. Replacing military standards and specifications with commercial standards and specifications increased the risks in these programs, because commercial standards and specifications did not meet the long mission life or take into account the need for nuclear hardening. The quality-affordability tradeoffs that characterize the commercial market do not necessarily fit military applications with their higher standards and lower tolerance for failure.

Eliminating lower-level tests (component-level and subsystem-level) often meant that faulty parts (e.g., resulting from a design flaw or a parts quality issue) were more likely to be discovered during system-level tests; this increased costs, because the satellite had to be taken apart, potentially down to the component level, to resolve the problem.

When underfunding, weakened oversight, and diminished government and contractor expertise were coupled with TSPR, the result proved problematic. Program costs were often underestimated under the assumption that new, albeit undemonstrated, efficiencies would result in lower program costs.[20] When these estimates were used as the basis of program budgets and the efficiencies were not realized, cost inevitably grew. Underfunding, in turn, created pressures to take even more risks, such as reducing systems engineering and testing, reducing subcontractor management, and so forth.

Although TSPR did not relieve DoD personnel of all of their responsibilities, it eroded their ability to exert them. DoD program managers were told to replace the detailed oversight of their contractors with a lighter touch emphasizing insight

---

[19] Hura et al., 2007.

[20] GAO, 2006; Younossi et al., 2008; Hura et al., 2007; SMC, 2013b.

into the companies they had previously overseen. The DoD acquisition and technical workforce shrank together with the rest of DoD in the 1990s (unrelated to the acquisition reform).[21] A shortfall of technical expertise kept the workforce from managing all phases of an acquisition program well. Contractors' technical expertise also fell with the budget-induced decline in the number of defense contractors able to serve as prime contractors or major subcontractors.[22] A similar erosion of expertise in the supplier base was cited as one key contributor to workmanship and design errors.[23]

Ultimately, these risks cascaded into design and workmanship errors, integration issues, and parts quality problems, which led to costly redesign, rework, and additional testing.

Commercial Space Market

Another unwanted surprise for DoD space acquisition arose from the overly optimistic forecasts about the commercial space market. In the early 1990s, growing awareness of the information revolution coupled with globalization convinced many investors that the prospects for global communications and hence the space industry were very bright. One such investment was Iridium, a network of 66 satellites (plus spares) that would provide telephone services for global businesspeople, particularly in underdeveloped countries with weak infrastructures. Globalstar, with 48 satellites, aimed at a similar market. An even more ambitious project was Teledesic, which was envisioned as an 840-satellite constellation. All three were slated for low-earth orbit. Until that point, almost all commercial communications satellites were in high-earth orbit, and even so, there were many new proposals, such as Lockheed Martin's Astrolink (operating in the Ka-band) and Hughes Aircraft's Spaceway (a "superhighway in the sky," also in the Ka-band).

Few of these dreams matched the hopes of their founders. Iridium and Globalstar were built, but both companies fell into bankruptcy. Both constellations were sold for pennies on the dollar. Teledesic was abandoned before anything was launched. Astrolink suffered the same fate. The Spaceway system was partially built, but as a North American rather than a global system and without satellite-to-satellite links.

The anticipated growth of commercial space industry did not materialize because of economic factors. Satellites were generally uncompetitive with fiber optic lines (for

---

[21] DoD had already reduced its acquisition workforce by over 30 percent from the end of FY 1990 to the end of FY 1995. Then, Congress, in its National Defense Authorization Act for FY 1996, mandated further reductions in DoD's acquisition workforce. The reductions led to departure of skilled, experienced scientists, engineers, and managers who played an essential role in DoD programs (Hura et al., 2007).

[22] Hura et al., 2007.

[23] McKinsey & Co, 2012.

global transmission) and cell phone towers (for local distribution). Further, the rapid spread of the cell phone infrastructure meant that hitherto dark spots on the world's communications grid were lighting up—and the attractiveness of paying a premium for the option to make calls from such locations was growing correspondingly less attractive.

The implications for military space systems were indirect but telling. In a world of burgeoning commercial satellite production, many were convinced that DoD's needs could ride on the margins of the large base that commercial manufacturing would provide. This perception may have helped convince DoD leaders that a commercially oriented acquisition policy could work. By then, the mental model of commercial technology as a spinoff from military work had mutated into a model of defense technology as a spinoff from commercial work[24] (to be fair, this impression arose from contemplation of the information technology, not the satellite, industry).

The most obvious victim of this misperception was WGS.[25] Boeing's initial low (and, ultimately unsustainable) bid was predicated on the assumption that Boeing (the new owner of Hughes Aircraft) would be able to amortize its factories and skilled manpower over a large business base.[26] When these hopes were dashed, the WGS program became more of a mainstay than add-on business. More of Boeing's costs had to be allocated to it (the alternative being to walk away from building satellites altogether), thereby raising bids on follow-on buys.

Programs to buy satellites with more specialized payloads and hence military specifications were less likely to have been affected by such optimism, largely because the parts-makers for military satellites were less likely to be commercial, but expectations of having fuller factories against which to write off the costs of handling assembly integration and test for military satellites may have also artificially depressed bid prices.

Summary

The end of the Cold War profoundly changed U.S. national security strategy and the defense budget in the 1990s, when many current space programs started. Such changes led to new requirements for increased tactical support from space programs, a reduction in both the government and contractor workforce, and acquisition reform that focused on cost and shifted to commercial standards, products, and practices. However, the resulting high-risk acquisition approach was plagued by complex and ambitious requirements; abdication of government responsibility and key acquisition

---

[24] See, for instance, Alic et al., 1992.

[25] The EELV program was another that incurred a significant cost growth as a result of the commercial space market not materializing (GAO, 2006).

[26] Blickstein et al., 2011.

and program management principles, such as strong systems engineering and test as a result of overaggressive implementation of TSPR; and overly optimistic assumptions about the commercial market and applicability of commercial practices, standards, and products for military space systems. Ultimately, the high-risk approach contributed to integration issues, workmanship and design errors, and parts problems, which required costly redesign, rework, and additional testing.

CHAPTER FOUR

Space Acquisition Challenges and Space Enterprise Management Issues

DoD space acquisition faces many inherent challenges that, if not well managed, can contribute to acquisition inefficiencies and difficulties. We begin this chapter with a discussion of these challenges common to all DoD space acquisition programs. We then discuss key space enterprise management issues and the effects they have had on program decisions and acquisition difficulties. These space enterprise management issues are in system of systems integration (i.e., synchronization of satellites, launchers, ground control and user equipment, and constellation management), and risk posture.

Space Acquisition Challenges

Space acquisition faces many challenges that can contribute to cost growth, schedule delays, and technical issues if they are not well managed. They include low-quantity buys, the limited industrial base, very stringent standards for components (e.g., space-qualified), high technological complexity, and inability to repair hardware on-orbit cost-effectively.

Low-Quantity Buys

Space systems, as with ships and early UAVs, differ from other procurements in that they are purchased in small numbers. Although the National Aeronautics and Space Administration (NASA) and the National Reconnaissance Office (NRO) also buy satellites, both buy systems in quantities too small to make any substantial difference in the total number of government satellites being purchased. The military satellites that DoD acquired had limited commonalities with commercial satellites to mitigate the low-volume problem, partially because military-unique systems, such as missile warning satellites and nuclear-hardened communications satellites, do not have a commercial market. This is not to say that opportunities to leverage the commercial market do not exist. Commercial satellites are used for military purposes, such as in providing commercial wideband SATCOM services, and commercial imagery data. Use of commercial buses may also be able to offer some benefit if they meet military specifications and standards and do not require significant modifications to support mission-unique requirements.

Purchasing in low volumes has several ramifications. First, there is little learning curve benefit. The last of the production run, although less costly than the first, will not have the cost reduction that accompanies large runs. Second, low volumes make it hard to break out all or even major parts of the design for competition or second-sourcing as a way to reduce cost, because the learning cost of a new contractor cannot be absorbed in a small production run. Third, contractors have few incentives to invest in ways to cut costs and schedules, given the difficulties in returning enough savings to justify such investments.

Unstable production schedules exacerbate the effects of low-volume production. As Chapter Two notes, many DoD programs have experienced multiple changes in buy quantity and long gaps between orders. Space systems are commonly purchased one or two at a time (partly because of the high unit cost of a space system) even though larger quantity buys could offer substantial cost savings.[1] The production gaps between satellites for the programs we reviewed are presented in Figure 4.1. Many of the programs have experienced a production gap greater than two years and, in the case of SBIRS, roughly eight years. Start-stop schedules lead to the flattening of learning curves, the

Figure 4.1
Long Gaps Between Orders and Unpredictable Buy Schedule

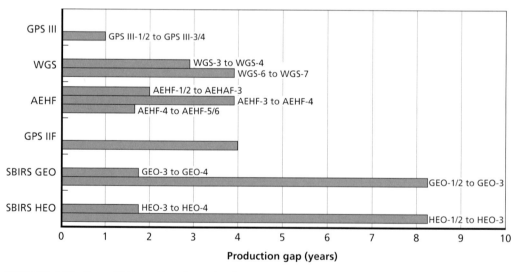

SOURCES: SARs; Hura, 2011; Lockheed Martin. 2012a.
RAND MG1171/7-4.1

---

[1]   High unit cost is one factor that limits government's ability to implement efficient buying practices, because of the near-term budget constraint, even though a larger quantity buy would be more cost-effective over the long term.

inefficient use of labor, and small-lot material buys. The longer it takes to complete a satellite program, the more redesign and nonrecurring engineering costs arise when parts go out of production. Instability in the production line has contributed to cost and schedule growth in later purchases of SBIRS, AEHF, and WGS.

Limited Industrial Base

Buying satellite systems a few at a time has affected the size of the space industrial base much as similar practices have hindered the shipbuilding industrial base; both complicate fostering and sustaining a competitive environment to promote contractor efficiencies and innovations. The space industrial base shrank significantly as a result of the significant defense industrial base consolidation in the 1990s.[2] About 70 aerospace contractors consolidated into five major contractors between 1990 and the early 2000s. The prime contractor of the current DoD satellite programs of records are either Boeing or Lockheed Martin.[3] Because Boeing focuses on communications and navigation space systems, Lockheed Martin is the only prime contractor with recent experience in space-based missile warning systems (Northrop Grumman, as a major subcontractor to Lockheed, provides the payloads for SBIRS and AEHF).

The high qualification standards required for space system parts have limited the growth of supplier base and competition, particularly when the qualification process itself requires special infrastructure. Quality requirements for government satellites can be more demanding than those for commercial satellites to ensure high reliability and hence long life for increasingly complex parts.[4] For example, GPS III experienced unanticipated cost growth because of the parts testing required when some lower-tier suppliers could not meet these requirements; the prime contractor had to bring parts testing in-house.[5]

Stringent Standards

Standards for components of space systems are high, and parts quality problems in satellites were common to all the space programs we examined in Chapter Two. Parts issues caused significant redesign and rework and, hence, schedule delays and cost growth in SBIRS, GPS IIF, and AEHF; for example, redesign and rework of the reaction wheel assembly (RWA) on AEHF added $200 million to the program's cost.[6]

---

[2] Hura et al., 2007.

[3] Orbital, Alliant Techsystems Inc. and Ball Aerospace are focused on smaller-sized satellites. Northrop Grumman does not have a commercial satellite business, and Space Systems Loral does not have national security space satellite business.

[4] GAO, 2011.

[5] "Parts Testing Drives Up GPS III Program Costs . . . ," 2012.

[6] Hura et al., 2011.

The RWA issue also hit the GPS IIF program, resulting in $10 million cost growth in redesign efforts.[7]

A 2011 GAO study found that inadequate parts quality, particularly electronics, in space programs was prevalent. All 21 programs assessed by GAO had quality issues originating in poor workmanship, design, and manufacturing practices. GAO attributed these practices to broader environmental issues, such as workforce gaps, diffuse leadership, the government's decreasing influence in the market for electronic parts, and a general increase in counterfeit parts.

High Technological Complexity

Complexity itself has contributed to satellite cost and schedule growth as shown in many studies.[8] Complexity in space systems may be attributed to the fact that space system hardware cannot be repaired once launched, which we will discuss below. But other complexities result from immature technologies, a high degree of integration with other complex components, subsystems, and systems, and complex failure modes. As a result, complex systems can be more difficult to integrate, model, and test than less-complex systems.

GAO's assessment of space acquisition over the last decade repeatedly found that technology was pushed too early in the DoD space acquisition cycle and programs suffered accordingly. Many of our interviewees echoed this sentiment. Using immature technology created problems in development of SBIRS' sensors, AEHF's cryptographic equipment, GPS IIF's M-code ASIC, and WGS's phased array. Past studies have found that immature technologies enter the acquisition process because acquisition is better funded than science and technology, there is a lack of independent assessments on when projects are actually mature,[9] and there are few technology insertion opportunities because of the lengthy development time lines.[10] Measuring technological readiness is far from easy. Although the technology readiness level (TRL) scale is used to estimate the technology maturity level, TRL is neither a forward-looking scale system nor a linear one. For instance, the time and resources it takes to go from TRL 3 to TRL 4 may not be the same as going from, say, TRL 6 to TRL 7. Further, the time it takes one technology to go from TRL 3 to TRL 4 can be very different for another technology, and the fact that a technology has gotten to any given TRL does not guarantee that it will ever get to a higher TRL.

---

[7]   GAO, 2009b. Several NASA satellites were experiencing anomalies with their RWA—the same ones used on SV-1 and SV-2.

[8]   For example, McKinsey & Co., 2012; Coonce et al., 2008.

[9]   McKinsey & Co, 2012.

[10]   GAO, 2007b.

Challenges Induced by the Inability to Repair On-Orbit Space Systems

A major source of complexity arises from the fact that repairing hardware on-orbit is nearly impossible (at least not cost-effectively),[11] and certain events, such as reaching orbit, do not permit a second attempt (e.g., because of launch failures).[12] Unforgiving circumstances coupled with the requirement that space systems must operate continuously over the entire life of the system means that high reliability is required from these systems. High reliability, in turn, requires high engineering margins and rigorous testing to survive the harsh launch environment (i.e., shock and vibration) and space environment (i.e., thermal, radiation, and vacuum). Preflight qualification tests must sort through every reasonable source of failure.

Designing and developing space systems is further complicated by tight spaces (measured in centimeters) available on a satellite for integration and the fact that ground testing cannot exactly replicate the operational environment. Experience has shown that some defects manifest themselves only in space (e.g., GPS IIF clock failure and the AEHF apogee engine failure), in which case the "repair" of spacecraft physical components is done through redundancy, switching from a failed component to an onboard replacement. GPS satellites carry three or more atomic clocks for that reason.

Redundancy and mission assurance (developed through engineering and testing) activities increase the cost of the spacecraft.[13] The high cost, in turn, fosters the desire for longer mission life, hence increasing engineering and testing to attain higher reliability, which further increases the cost of the system and the time required for development. The increase in spacecraft cost also has implications for the launch vehicle. The launch vehicle must also provide high reliability to ensure that the expensive spacecraft reaches its orbit, thereby driving launch vehicle costs (on the order of $10,000 per pound to low earth orbit[14]). The high launch cost further encourages longer space mission life to avoid incurring launch costs frequently, continuing the vicious circle between high reliability and high costs (Figure 4.2). Many interviewees also said that the high launch cost was one driver in aggregating multiple missions and requirements onto a single platform.

Long mission life requirements combined with the lengthy development make technology insertion difficult and foster the temptation to use the latest immature state-of-the-art technology. The program must anticipate requirements over a 20-year period (its terrestrial cousins can adapt by using in-service modification upgrades). As a

---

[11] Astronauts have conducted on-orbit repairs, but having a manned space program to repair satellites on-orbit would not be cost-effective. In the future, on-orbit hardware repair for some satellite components may be feasible as technologies in robotics and rendezvous and proximity operation advance.

[12] Although software can be patched remotely, the rising software content in space systems (as with all defense systems) accompanies increases in complexity.

[13] Northrop Grumman, 2013a.

[14] This estimate is based on the EELV cost. The launch price varies depending on the launch provider, the launch vehicle, and other mission-unique services that a customer may require.

Figure 4.2
Vicious Circle Between Satellite High Reliability and Satellite Cost

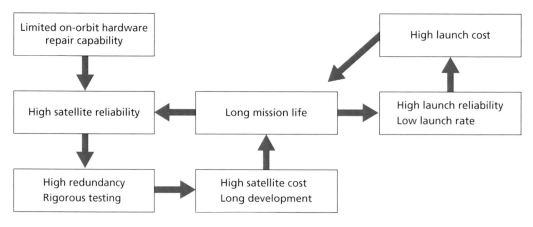

RAND *MG1171/7-4.2*

result, the requirements for DoD space systems are subject to high levels of uncertainty. In addition, because of the technical uncertainty that generally surrounds space-based systems, accurate cost estimates may be very difficult to achieve in new programs, especially early in the process. The long development cycle also introduces more opportunities for perturbations in the program (requirements changes, funding reduction, etc.), which generally have an adverse effect on the program.[15]

Space Enterprise Management Issues

System of Systems Challenges: Synchronization

Delivering space capabilities to the users requires synchronizing delivery of the satellites, the launch vehicles, the ground control systems (to command and control the satellite), and the user equipment (to receive data or signals). Interfaces among these space, ground, and user segment systems create additional system integration risks, and these could lead to additional costs and longer delivery times if not well managed. Synchronization of these segments is challenging because they are usually managed by different organizations.[16] The satellite program office is responsible for delivering the satellites and the ground control segment (although the GPS III program split the funding of the GPS III satellites from its ground control segment,), the EELV program office acquires the launch services, and the user equipment (e.g., the Air Force's Family

---

[15] SMC, 2013b.

[16] GAO, 2009d. The GAO study found that six of eight space systems had a hard time delivering ground and user assets at the same time as the launch of satellite systems.

of Advanced Beyond Line-of-Sight Terminals (FAB-T), the Army's Warfighter Information Network–Tactical) is the responsibility of the individual military services.[17] Many of these programs are individual large-scale ACAT 1D programs (e.g., satellites, Operational Control Segment, EELV, and FAB-T), which further adds complexity to overall system architecture integration and synchronization.

Table 4.1 highlights some of the challenges for DoD satellite programs in synchronizing with the launch vehicles. Changes in the launch vehicle may induce changes in the satellite design, because the vibration and velocity profiles as well as the physical satellite support points differ significantly for each launch vehicle.[18] Both SBIRS and AEHF satellites had to reduce their weight to be compatible with the less-expensive, medium-class launch vehicle they were supposed to use rather than the heavy-class launch vehicle their predecessors, DSP and Milstar, used.[19] This reduction in satel-

Table 4.1
Examples of Space and Launch Segment Synchronization Challenges

Space Program	Challenge
SBIRS	SBIRS was under pressure to reduce weight (compared to its predecessor DSP) to be able to launch on a less-expensive, medium-class launch vehicle while increasing satellite capability, which further increased complexity.
GPS IIF	When spacecraft weight increased as a result of "modernization," a larger launch vehicle was needed. Throughput issues at the ranges (all launching from Eastern Range) could lead to potential launch delays.
AEHF	AEHF was under pressure to reduce satellite weight (compared to its predecessor Milstar) to be able to launch on a less-expensive, medium-class launch vehicle while increasing satellite capability, which further increased complexity. AEHF weight growth led to changing to a larger launch vehicle, an intermediate-class vehicle, leading to cost growth.
WGS	Boeing added extra solar panels to the original design, which added weight to the satellite, causing changes in the launch vehicle for both Delta IV and Atlas V.[a] WGS-5 was delayed five months because of an issue with Delta IV's upper stage RL10B engine.[b]
GPS III	Launch may be delayed because of a lack of launch vehicle availability.

[a] The first WGS satellite was supposed to be launched on Delta IV and the second WGS on Atlas V (DOT&E, undated [a]) but WGS-1 and WGS-2 both launched on Atlas V in 2008 and 2009, respectively.
[b] Gruss, 2013.

[17] Each service is managing modernization of its GPS user equipment through the joint Military GPS User Equipment program.

[18] Northrop Grumman, 2013a. Conducting a coupled loads analysis could be a multiple-year activity involving analysis of the launch vehicle environment and loads to ensure that the satellites are compatible with the launch environment; this may require satellite redesign to ensure that the satellites can withstand the launch environment.

[19] Hura et al., 2007, 2011.

lite weight had to be accomplished while accommodating the substantial capability increase. Conversely, any increase in the satellite weight or volume could result in changing the launch vehicle in midstream, which was the case with AEHF, WGS, and GPS IIF.[20] These changes lead to additional work, potential contract modifications, and, ultimately, schedule delays and cost growth.

Some satellite launches may be delayed when launch vehicles are not ready or there is no room on the schedule because of limited throughput capacity at the launch sites. Storing satellites awaiting their launch also adds costs. Many DoD satellites (SBIRS, GPS IIF, AEHF, and WGS) were delivered later than originally scheduled because of their development difficulties. These satellites started to be delivered around 2010, and this may have created congestion at the ranges (especially in the Eastern Range), because of the limited launch throughput capacity. For example, eight GPS IIF satellites (SV-2 through SV-9) were delivered between March 2011 and December 2012. With the current throughput capacity at about 10–12 launches per year and other satellites competing for the limited launch opportunities, it would have been difficult to launch all eight GPS IIF satellites shortly after delivery. In fact, GPS IIF satellites SV-3 through SV-7 spent an extended period in storage, from 18 months to 26 months, while waiting for their launch.[21] GPS IIF satellite SV-8 is scheduled to be launched in October 2014, about 22 months after its delivery, and GPS IIF satellites SV-9 through SV-12, which were delivered in 2013, are planned to be launched between 2015 and 2016.

Ground control system development and synchronization problems can also hinder satellite acquisition. Ground control stations are, themselves, complex, and they usually support only a single type of satellite. Multiple satellite programs, such as GPS, SBIRS, WGS, and AEHF, were affected when their ground control segments fell behind the satellite schedule, which led to delays in space vehicle launch schedules (GPS III), delays in satellite operational testing (SBIRS, WGS, and AEHF), and unusable capabilities on the existing satellites (GPS IIF, SBIRS, AEHF, WGS, and GPS III).[22] Similarly, lack of synchronization with the user equipment also caused delays in delivery of capability to warfighters and affected the operational testing of the satellites and causing delays.

Table 4.2 summarizes the gap between satellite launch, the deployment of ground control systems, and the fielding of user equipment. The SBIRS program experienced problems with its ground control segment early in the program. Although the first SBIRS GEO satellite was launched in May 2011 with both scanner and starer sensors, the ground segment software to process the starer sensor's data was not scheduled to

---

[20] GAO, 2003a, 2006; DOT&E, undated (a).

[21] Based on the GPS IIF satellites delivery dates reported in the NAVSTAR SARs (2011–2013) and the GPS IIF satellites launch dates announced in open sources.

[22] GAO, 2009d.

Table 4.2
Gap Between Fielding of Satellites and Their Ground Control and User Segments

Program		Key Milestone	Gap
SBIRS	Ground control system	First SBIRS GEO launch in May 2011 Delivery of full Ground Increment 2 in 2018	7 years
	User equipment	N/A	N/A
GPS IIF	Ground control segment: OCS	First GPS IIF launch in May 2010 Delivery of OCX Block I in 2016 Delivery of OCX Block II in 2017	6–7 years
	User equipment	New military signal IOC in 2015–2016 time frame (estimated) Fielding of modernized user GPS equipment estimated to be complete in 2025	10 years
AEHF	Ground control system	First AEHF launch in August 2010 Delivery of Mission Control System in June 2015	5 years
	User equipment	AEHF IOC in June 2015 Air Force terminal FAB-T IOC in 2019	4 years
WGS	Ground control system	First WGS launch in October 2007 IOC declared in January 2009, but mission planning system (Common Network Planning Software) did not work as intended	15 months
	User equipment	First WGS launch in October 2007 Predator and Reaper were equipped with commercial SATCOM terminals because of long delays in the first WGS satellite; these unmanned aerial systems may be equipped with WGS terminals in the future	N/A
GPS III	Ground control system	First GPS III launch 2015 Delivery of OCX Block I in 2016 Delivery of OCX Block II in 2017	1–2 years
	User equipment	New military signal IOC in 2015–2016 time frame (estimated) Complete fielding of modernized user GPS equipment in 2025	10 years

SOURCES: GAO, 2009b, 2009d, 2012, 2013b; Butler, 2013; "Air Force Commander: GPS III, OCX Delayed," 2012.

be completed until 2018.[23] The ground segment delivery schedule was subsequently altered to be incremental, providing some of the otherwise unavailable capabilities at earlier dates, but the full suite of usable data from the "critical sensor" will still not be acquired before 2018.[24]

---

[23] GAO, 2012.

[24] GAO, 2013b.

Similarly, despite the launch of three AEHF satellites so far, its IOC will be delayed to 2015 because of software issues in its ground control segment.[25] Because the Air Force's FAB-T program (which provides AEHF user terminals for airborne platforms) is late, AEHF capability cannot be fully used by many platforms until at least 2019.[26]

The ground segment for WGS also lagged behind the first launch of the WGS satellite. The mission planning system for WGS, the Common Network Planning Software, did not function as intended when WGS attained its IOC.[27] It was acquired separately from the satellite, and its shortcomings affect users' ability to plan WGS communications. Conversely, the four-year delay in the WGS Block I satellite forced the Army and the Air Force to acquire commercial satellite terminals for some of their systems and the Predator and Reaper unmanned aerial systems, respectively.[28]

Delivery of the GPS IIR-M and GPS IIF ground control segment, OCS, when their respective satellites were ready, has been difficult—partly because OCS funding was diverted to the GPS IIF satellite development effort resulting in the deferral of OCS capabilities to the next upgrade in the GPS control segment.[29] Specifically, the new civil signal (L2C), the new military signal (M-code) on IIR-M and IIF, and the third civil signal (L5) on IIF will be unusable until OCX Block I and Block II are delivered (roughly 2016–2017).[30] This schedule represents a three-year delay.[31]

OCX development proved more complex than originally anticipated, in part because of such emerging issues as cybersecurity.[32] The decision to develop the OCX as a separate program from GPS III was supposed to help the OCX program in three ways.[33] First, it allowed a competition for the ground segment, in hopes that the government could get a better contract than if the winner of the satellite were automati-

---

[25] SAR, December 2012.

[26] GAO, 2013b.

[27] GAO, 2009d. As of October 2009, the mission planning system issue was not resolved, according to a GAO report (GAO, 2009d).

[28] SMC, 2013b; Butler, 2013.

[29] GAO, 2009b. Other difficulties in OCS included coordinating OCS development for IIR and IIF (Younossi et al., 2008).

[30] The new civil signal (L2C) on IIR-M and IIF will be unusable until OCX Block I is available, and new military signal (M-code) on IIR-M and IIF and the third civil signal (L5) on IIF will not be operational until OCX Block II is available (SAR, December 2012; National Coordination Office for Space-Based Positioning, Navigation, and Timing, 2013).

[31] GAO, 2009b.

[32] Starosta, 2012.

[33] Unlike with previous GPS programs, the Air Force opted to develop the satellite and the corresponding ground system as separate programs after 2005 (SAR, 2008).

cally the developer of OCX.[34] Second, it permitted OCX to start earlier, inasmuch as the satellite contract (expected in 2007) slipped.[35] Third, it prevented the common practice of shifting ground segment funds to make up for satellite cost overruns.[36] But separation meant that plans for both segments were disconnected, leading to major delays in harnessing existing satellite capabilities and problems with oversight leading to GAO findings. Recognizing significant synchronization issues between GPS satellites, ground stations, and user equipment, OUSD (AT&L) established the Annual GPS Enterprise Review in 2009 for an integrated enterprise-level review of GPS capabilities, including synchronization risks between the various GPS segments and segment increments to support the decisionmaking process.[37]

Although the first GPS III satellite was to be launched in 2014, the date slipped to 2015 so that OCX Block 0 would be available to help perform on-orbit checkout of the satellite.[38] Many of the IIR-M, IIF, and III satellite capabilities will remain unusable until the later blocks of OCX are delivered.[39]

The delay in the ground systems interfered with the operational testing of the satellites and limited the opportunity to improve or adjust later satellites. In the case of GPS, all of the IIR-M satellites have already launched, and all of IIF satellites will be launched by the time OCX is available to test the new capabilities on these satellites. The fielding of modernized GPS user equipment to receive and process the modernized signal on GPS satellites may not be complete until as late as ten years after the military signal becomes initially operational.[40] For SBIRS, four SBIRS GEO satellites will have been launched by the time a fully operational ground segment is available.

System of Systems Challenges: Constellation Management

Ensuring that there are no gaps between legacy and replacement satellites entails solving complex constellation management challenges, such as the timing of the deployment of replacement satellites and their backward-compatibility with legacy systems. For example, GPS consists of GPS IIA, GPS IIR, GPS IIR-M, GPS IIF, and, soon, GPS III—a sufficient population of which must be in orbit to meet the minimum constellation size to support the mission lest there be inadequate regional coverage or

---

[34] Simpson, 2008b.

[35] Simpson, 2008a.

[36] GAO, 2009d.

[37] Carter, 2009.

[38] "Air Force Commander: GPS III, OCX Delayed," 2012.

[39] GAO, 2012.

[40] The IOC of the new military signal is reached when at least 18 satellites can broadcast the signal. Eight GPS IIR-M with the new military signal are already on-orbit and the remaining four GPS IIF satellites are planned to be launched by 2016 (National Coordination Office for Space-Based Positioning, Navigation, and Timing, 2014); and the user equipment is expected to be fully fielded by 2025 (GAO, 2012).

reduced accuracy. Unlike terrestrial, airborne, and sea-based systems whose service life can often be extended if recapitalization is delayed, there is no way to make up shortfalls except with a replacement satellite; absent spares, recovering from failure could take years. GPS is an exception, because 24 operational satellites are backed up with seven on-orbit spares, although concerns about operational availability have been raised in recent years because of the potentially shortened life of some GPS IIF satellites.[41] Hence, replacement satellites must be ready to launch before the legacy system ages and fails to ensure continuity of mission.

Program schedules, driven by the need for mission continuity, can be highly vulnerable to surprises (e.g., launch or satellite failures). The programs studied in our research all experienced schedule pressure throughout their programs, arising from the following:

- aging legacy systems
- early termination of production of the legacy system (before adequately understanding the risk in the replacement program)
- unanticipated failure(s) in the constellation
- unanticipated gap in the constellation as a result of a change in the constellation mix or a new requirement
- schedule delays because of program difficulties.

These factors are summarized in Table 4.3. Many DoD space programs experienced difficulties—sometimes as a result of attempted schedule acceleration—that led to delays, which in turn created more schedule pressure to avoid an operational gap. Schedule pressures ultimately reduced schedule margins and fostered a risky path to acquisition, such as a reduction in testing and beginning programs before requirements are well defined or critical technologies are mature.

Shift in Risk Posture

Some of the challenges of space acquisition are new (e.g., limited industrial base, limited quantity buys, reduced acquisition workforce). Others are old (inability to repair hardware on-orbit, system of systems complexities) but they were managed differently in the past.[42] Before the 1980s, government was the primary customer of all space systems and the driver of their development. Because national survival was at stake, space system development and fielding was a top national priority. The space industrial base was growing, and the business environment was competitive. The urgency of requirements meant that funding necessary for the successful completion of programs

---

[41] The GPS constellation requires 24 satellites to be available 95 percent of the time (National Coordination Office for Space-Based Positioning, Navigation, and Timing, 2013).

[42] Hura et al., 2007.

Table 4.3
Factors That Contributed to Schedule Pressures in DoD Space Programs

Space Program	Factor
SBIRS	DSP-24 and DSP-25 were cancelled in 1994 DSP was nearing the end of life and SBIRS began an accelerated path in 1996 DSP-19 failed to reach orbit in 1999 Continued delays in delivery of first SBIRS GEO satellite (from originally planned 2002 launch to actual launch in 2011)
AEHF	Aging Milstar Milstar flight 3 failure in 1999 created a gap in the constellation AEHF began with 18-month acceleration TSAT needed for AEHF to reach full operational capability but TSAT was cancelled in 2007, creating a gap in constellation Continued delays in delivery of first AEHF (from originally planned 2004 launch to actual launch in 2010)
GPS IIF	Aging GPS IIAs and GPS IIRs Three-year gap in program while modernization plan was being developed Continued delays in delivery of first GPS IIF (from originally planned 2005 launch to actual launch in 2010)
WGS	Aging DSCS Anticipated gap in wideband MILSATCOM as a result of accelerated growth in requirement TSAT cancellation created a gap in the constellation Continued delays in delivery of first WGS (from originally planned 2004 launch to actual launch in 2007)
GPS III	Aging IIR-M and GPS IIF Truncated GPS IIF at 12 satellites vs. 33 satellites Concerns about GPS IIF's remaining life being shortened drove the need to launch GPS III by FY 2013 GPS III program start delayed by six months

SOURCES: GAO, 1997; "Defense Support Program Satellite Decommissioned," 2008; DOT&E, undated (b); Moody, 1997.

was provided expeditiously, which created a tolerance for program cost growth. It also encouraged the education of government experts in space acquisition management, science, and engineering. The proficiency of contractor scientists, engineers, and acquisition management personnel also increased.

When space was a new domain, the development and fielding of space capabilities was exploratory. Failures were tolerated, as Table 4.4 indicates by contrasting the failure-tolerant approach to space acquisition then with today's much more failure-intolerant approach. Uncertainties about the reliability of space systems then meant short design-life requirements ranging from months to a few years. Consequently, the development cycle for these systems was short, about five years.[43] With government concurrence, contractors largely followed an evolutionary acquisition strategy, developing demonstration articles and prototypes before production models to get a better understanding of the technology, and inserting technology upgrades to follow-on sys-

---

[43] Northrop Grumman, 2013a.

Table 4.4
Shift in Risk Posture from Early Space Programs to Current Space Programs

Early Space Programs (Failure-Tolerant)	Current Space Programs (Failure-Intolerant)
Uncertainty about satellite reliability, hence short expected mission life (months to a few years)	Long mission life is expected and demanded (>12 years)
Prototypes built to test/demonstrate technology	Satellites must work the first time; test article = operational article
Launch on schedule (leading to spare satellites on orbit and on ground in storage)	Launch on need to get maximum life of each satellite (leading to no spare satellites on orbit or on the ground in storage)[a]
Production of legacy system while replacement system is being developed	Minimal overlap between existing and future systems

[a] GPS constellation is an exception.

tems incrementally based on lessons learned from previous increments. In parallel, government and contractor personnel had to develop engineering and manufacturing processes, procedures, and capabilities to produce space parts, components, subsystems, systems, and mission architectures. To ensure that there was no operational gap in the event of a launch or an on-orbit satellite failure, satellites were launched on schedule. Over time, some of these satellites outlived their design life, leading to excess satellites on-orbit and on the ground available as spares in case of an on-orbit or launch failure. Additionally, the legacy systems were in production while the new replacement systems were being developed.

Maturation of the space industry led to a failure-intolerant approach to acquisition. As spacecraft became more reliable and lasted much longer, a longer life was expected as a requirement: 12 to 15 years typifies the five programs we examined. The design lives of SBIRS, GPS IIF, AEHF, WGS, and GPS III are 12, 12, 14, 14, and 15 years, respectively.[44] Today, the first development satellite built is the first operational satellite that must work to near-perfection to meet the long mission life requirement. This requirement creates a tremendous pressure to ensure that the first development satellite—a highly expensive system—is, in fact, highly reliable. There are no longer prototypes to enable learning, leaving that much less room for error.[45]

Furthermore, to save costs, there are no on-orbit or stored spare satellites (except for GPS); if unanticipated failure occurs, recovery time can be a few years. Further, the overlap between on-orbit systems and their replacement is reduced, creating yet more pressure to ensure that each satellite is highly reliable and launched on schedule.

---

[44] Hura et al., 2007; Jackson, 2012; Trauberman, 2013.

[45] The GPS III program, which was started as a "back-to-basics" program, does have a testbed article.

Summary

Space acquisition faces many challenges that can contribute to cost growth, schedule delays, and technical issues if they are not well managed. They include low-quantity buys, the limited industrial base, very stringent standards for components (e.g., space-qualified), high technological complexity, and demand for high reliability. Given these inherent challenges, the high-risk acquisition approach taken by the programs as described in Chapter Two was not appropriate for space acquisition.

Several enterprise-level systemic issues further contributed to the adverse outcome of the programs. The long development cycles created their own risks by generating more opportunities for requirements changes, funding reductions, obsolescence, or unanticipated technical problems. Long cycles also fostered the temptation to use the latest, immature state-of-the-art technology. Lack of synchronization between all the segments in the space enterprise (satellite, launcher, ground control, and user equipment) and poor constellation management contributed to the inefficient buying practices, acceleration of program schedules, and unanticipated changes. Disparate management of the segments may have contributed to the difficulties in synchronization. Finally, the failure-intolerant risk posture combined with the goal of optimizing each satellite's utility (e.g., maximum use of satellite service life, using the "test" article as an operational system) allowed very little room for error for these long, complex, development programs. There was minimal margin (schedule, technical, and cost) to deal with unanticipated problems, making space acquisition susceptible to cost growth and schedule overruns.

CHAPTER FIVE

Recent Progress and Future Challenges in DoD Space Acquisition

This chapter examines the current state of space acquisition. We highlight changes in acquisition policies and practices that address earlier space acquisition challenges. We then discuss the challenges for next-generation space systems and their implications for space acquisition.

Recent Progress in Space Acquisition

Progress is being made in space acquisition. Other than the GPS IIF program, which delivered all 12 GPS IIF satellites, the programs we examined are either in production or transitioning into the production phase.[1] Satellites are being launched, and cost and schedule growth appear to be stabilizing. This stability comes in part from getting through development and into production as technologies are matured and risks reduced.

Back to Basics and GPS III

As mentioned in Chapter Two, a back-to-basics acquisition approach is characterized by a block approach to delivering capabilities incrementally and sound foundations of systems engineering, business management, and risk management principles. Several interviewees felt that instituting a back-to-basics acquisition approach has been the most important improvement in space acquisition.

The GPS III program was the first DoD space program since the acquisition reform to follow the back-to-basics acquisition approach from inception.[2] To stabilize requirements—which is what drives system engineering—a senior council of users chaired by OUSD (AT&L) and the Under Secretary for Policy within the Department of Transportation was established to review progress and tasked to keep requirements stable (and if they could not be, the council would help find funds and adjust sched-

---

[1] As of fall 2013, GPS III-1 is in the final stages of development integration and testing and GPS III-2 is at the beginning of the production line (SMC, 2013b).

[2] TSAT also instituted a back-to-basics approach but was canceled before reaching Milestone B.

ules accordingly). In addition, the commander of Air Force Space Command was given authority to approve new requirements.[3]

The GPS III program also used a nonflying spacecraft, GNST, as a system engineering tool to identify and correct system integration problems in EMD and production as well as coordination issues between the satellite and other systems, such as receivers, ground control, and launch facility–handling equipment. The GNST helped transfer knowledge and expertise between the EMD facility in Pennsylvania and the assembly and testing facility in Colorado. The Systems Engineering Management Plan included a road map of system integration and risk reduction demonstrations. Modeling and simulation, as well as the use of a system integration lab, also played a large role. The results of these efforts modified the design and use of the GNST and the Test and Evaluation Master Plan.

The cost and schedule performance of GPS III, so far, has been notably better than the other four major DoD space programs that began in the acquisition reform period.[4] However, whether it stays on track remains to be seen. Its acquisition strategy,[5] timing of insertion of capability upgrades,[6] and possibly a different architecture (e.g., disaggregation) for GPS III follow-ons are all potential sources of risk.[7]

Efficiency and Cost-Saving Initiatives

The Air Force has recently implemented several efficiency and cost-saving initiatives that may reduce the cost of production satellites. The Air Force is working with contractors to identify ways to streamline the processing and production flow and adopting a block-buy approach for many programs.[8] Although such approaches look like the problematic reforms of the 1990s, the risks entailed in taking them are much less when satellites are in production rather than in development, because key technologies are mature and uncertainties are reduced when the program enters production.

One goal for streamlining the processing and production flow is to reduce the risks of schedule slippage.[9] For the acquisition of WGS-7 through WGS-10, the Air Force is allowing the prime contractor (Boeing) to use commercial standards and prac-

---

[3] Hura et al., 2011.

[4] Table 2.6 compares the program cost and schedule growth between GPS III and the other programs at a similar point in the acquisition cycle.

[5] SAF/AQS, 2013.

[6] Davis, 2013.

[7] GAO, 2013c.

[8] SMC, 2013b; Boeing, 2013; Lockheed, 2013a.

[9] SMC, 2013a; SMC, 2013b.

tices in systems engineering, test, and management, among other areas.[10] Again, these are risks that can be taken now that the design has been stabilized and the commercial SATCOM market is mature and stable (as it was not for WGS Block I, which was predicated on the unfulfilled promise of commercial satellite market growth). Because of this stability in the design and the market, government oversight is significantly reduced for the WGS Block II follow-on.

Additional saving is expected by implementing a block-buy approach with a fixed price contract for AEHF-5 and AEHF-6, SBIRS-5 and SBIRS-6, and WGS satellites SV-7 through SV-10.[11] Block buys will prevent costly production breaks, and stable funding fosters industrial base stability. The benefits require that DoD stay committed to the contract, but the current austere budget climate does create a risk of program cuts with implications for unit cost increases.[12] Table 5.1 summarizes the programs that have instituted the cost-saving measures discussed above.

The Air Force and the contractors (Lockheed Martin and Boeing) were able to reduce some testing, review, and reporting activities to reduce program costs such as the following:[13]

Table 5.1
Recent Efficiency and Cost-Saving Initiatives in Selected DoD Space Systems in Production

Space Program	Reduced Testing	Reduced Oversight	Lean Processing and Production Flow	Fixed Priced Contract or Block Buy
SBIRS	X	X	X	X
GPS IIF	X	X	X	X
AEHF	X	X	X	X
WGS	N/A[a]	X	N/A[a]	X
GPS III[b]	X	X	X	X

[a] Given the commercial-like acquisition model, we do not compare commercial processes with the government process.

[b] GPS III is transitioning into production. GPS III-1 is near completion of development testing and integration, and GPS III-2 entered the production line (in April 2013). GPS III is a cost-plus contract for the first four satellites. However, in May 2012, the GPS III program received approval to convert unexecuted space vehicle cost-plus options to fixed price incentive beginning with the fifth satellite (GPS III SAR, 2012).

---

[10] SMC, 2013b. It is unclear, though, how the commercial testing and production flow processes compare to those that the government would have specified.

[11] The WGS SV-9 satellite is being procured and funded separately under a cooperative agreement through an international partnership and MOU with Canada, Denmark, Luxembourg, the Netherlands, New Zealand, and the United States.

[12] McCullough, 2011.

[13] SMC, 2013b.

- reducing or eliminating tests based on information gathered from the development system and previous production systems
- reducing formal government/industry meetings and the number of meeting participants[14]
- reducing the number of contract data requirements.

Other issues need to be dealt with. The Air Force is introducing competition in the launch vehicle (EELV), user equipment (FAB-T), and sustainment activities (e.g., GPS control segment sustainment).[15] In the case of the EELV program, the threat of competition appears to have contributed to the launch cost reduction.[16] United Launch Alliance currently has a monopoly in national security space launches, but the Air Force is planning several competitive awards for national security space launch missions during 2015–2017. The introduction of competition for the follow-ons to the first block of GPS III satellites is also being considered. As of June 2014, the Air Force planned to award two fixed price contracts—the first for development of a navigation waveform design and the second, in the FY 2017 to FY 2018 time frame, for a limited competition between Lockheed Martin and an alternative contractor for up to 22 GPS III satellites.[17]

Competition may not always lead to a successful outcome, however. For example, the GPS III OCX program was separated from the space segment to introduce more competition, which led to the problems discussed in Chapter Four. There are broader concerns about whether the acquisition workforce has enough experience, hence expertise, to know when competition makes sense and how best to apply competition in various circumstances.[18] Although DoD has published guidance on promoting effective competition in its memo on better buying power 2.0,[19] its implementation directive also notes that a qualified acquisition workforce is a key overarching principle that underlies BBP and that "policies and process are of little use without acquisition professionals who are experienced."[20]

Although recent efficiency and cost-saving initiatives may reduce the cost of space systems, such systems remain expensive (e.g., more than $1 billion each). This points to the need for finding technology-based paths to affordability to address this issue.

---

[14] SMC developed the Program Operating Plan to establish protocols for the interaction between SMC and the contractor (SMC, 2012).

[15] Davis, 2013; Mehta, 2012; Host, 2012; SMC, 2013b.

[16] Gruss, 2014.

[17] Peck, 2014.

[18] SMC, 2013b.

[19] OUSD (AT&L), 2012.

[20] OUSD (AT&L), 2013a.

Technology Development

Air Force–sponsored technology development activities have been initiated to accelerate the development of future space systems, lower their cost, permit competition in their acquisition, and make their architecture better able to adapt to new threats. These include hosted payload concepts and smaller space systems to enable disaggregated mission architectures.

The Commercially Hosted Nonimaging Infrared Program developed and launched a wide field-of-view infrared (IR) sensor hosted on a commercial SATCOM system. In addition to demonstrating the military utility of this technology for missile warning capability, it pioneered the hosting of a DoD payload on a commercial bus, enabling faster (three years from contract award to launch) and cheaper acquisition, as well as possibly increasing the space architecture's resilience.[21] After the formation of the Industry Hosted Payload Alliance to explore other opportunities for commercially hosted payloads, SMC established a Hosted Payload office to synchronize its needs with the needs of payload- and host-makers.[22]

But hosted payloads are no panacea; they reduce costs only if the hosted payload satisfies the weight, space, and power constraints imposed by the commercial bus and if there is no electromagnetic inference between host and payload. The hosted payload provider must accept that the primary payload mission will always take precedence if there is a mission conflict.

The Air Force's Space Modernization Initiative is pursuing new technology to enable simpler, cheaper satellites to fit different space architectures. For example, smaller, low-cost PNT satellites to complement the core GPS III satellites are being explored.[23] For the protected MILSATCOM mission, the Space Modernization Initiative calls for technology development in the following areas:

- smaller, simpler, and more affordable National Command and Control capability evolved from the current AEHF system
- a dedicated tactical protected payload with multiple flight options
- simpler, lower-cost-protected communication terminal for new missions.[24]

---

[21] In June 2008, an $82.5 million contract (which included launch and operations) was awarded for the Commercially Hosted Nonimaging Infrared Program, with payload delivery in July 2009 and an expected launch on a commercial launch vehicle in May 2010; actual launch was in September 2011 (Aerospace Corporation, 2013).

[22] Foust, 2012.

[23] U.S. Air Force, 2012b. The funding for such demonstration (demonstration program called NavSat) remains uncertain.

[24] U.S. Air Force, 2012a.

Future Challenges and Potential Implications for Future Space Acquisition Programs

U.S. budgets are falling as other countries and even nonstate actors are developing capabilities to deny or disrupt space-based capabilities. This threat combination underscores the need for the current DoD space enterprise to counter its challenges more effectively.[25]

In 2007, China launched a missile to destroy a redundant Chinese weather satellite. Insofar as any country capable of putting objects into orbit can interfere with satellites in similarly distant orbits, such options may become available to other countries and actors of concern as well. Such countries are also working on using electronic warfare to jam uplinks to render communications satellites inoperable.

Concern has also arisen that cyberattacks on satellite ground control stations could hijack the functionality of satellites and perhaps alter their orbits enough to make them unrecoverable.[26] With growing cyberthreats, cybersecurity standards are tightening. Furthermore, the addition of the cyberwar community to the list of communities that want space systems to reflect their needs adds one more complication to the lives of program managers.

The last few years have seen growing interest in disaggregated architectures, which involve building satellite constellations from many smaller satellites rather than a few large ones as one of many approaches to addressing the emerging threats.[27] One disaggregated architecture concept that has been gaining attention is a constellation with many smaller satellites (including hosted payloads on commercial satellites) that are less complex and have a shorter mission life.[28] *Conceptually*, such a concept would address the following risk and complexity factors (discussed in Chapter Three) that contributed to past space acquisition difficulties:

- **Technology maturity:** Shorter mission life enables more frequent technology insertion opportunities, which could curb the temptation to insert the latest state-

---

[25] Pawlikowski, Loverro, and Cristler, 2012; AFSPC, 2013.

[26] The U.S.-China Economic and Security Review Commission reported in 2011 that hackers, possibly Chinese, had managed to control a Landsat for several minutes at a time by hacking into a Norwegian ground station (pp. 215–216); the information was largely based on a briefing that the U.S. Air Force provided to the commission on May 12, 2011.

[27] Our interviews indicated that the vision for the degree of disaggregation is not as extreme as was envisioned under the Defense Advanced Research Projects Agency F (Fractionated) 6 approach "wherein the functionality of a traditional monolithic spacecraft is delivered by a cluster of wirelessly-interconnected modules capable of sharing their resources and utilizing resources found elsewhere in the cluster" (Defense Advanced Research Projects Agency, undated; Ferster, 2013).

[28] Pawlikowski, Loverro, and Cristler, 2012.

of-the-art immature technology and thereby increase risk. A lower unit cost may also enable use of prototypes for risk reduction.

- **Low volume production:** A lower unit cost may enable DoD to purchase satellites in bulk, reducing costs, eliminating production gaps, and stabilizing the industrial base while allowing increased competition and thereby broadening DoD's access to the smaller satellite manufacturers.

- **Schedule pressure:** A lower unit cost could allow having spare systems or sufficient overlap between the legacy and replacement systems. This approach would reduce the pressure to accelerate programs and create a schedule buffer against such unanticipated problems as on-orbit failures.

- **Failure intolerance:** A lower unit cost, a shorter mission life, and a larger constellation may enable a more failure-tolerant approach to space acquisition, similar to that taken in the early space programs (see Table 4.4 and the associated text), which may allow more margins and flexibility for space program.

However, disaggregation exacerbates the complexity arising from the highly interdependent nature of the space enterprise. For instance, diversity in disaggregated architectures could further complicate synchronization and constellation management challenges. It might mean higher launch rates that may run into capacity bottlenecks at launch ranges or compel the construction of new ones. The high launch cost is also a significant concern, as discussed in Chapter Four. Command and control of satellites may also become more complex.[29] Further, there may be nonrecurring engineering components in all four segments (space, launch, ground control, and user), which could be significant, meaning that hoped-for economies of scale may be questionable.

Lessons Learned from the 1990s

In the post–Cold War era, an attempt to meet the increased tactical requirements in an affordable manner led to multimission systems with complex sets of requirements, which contributed to various difficulties. The future environment is creating a similar strain on space acquisition. Responding to emerging threats entails increased resilience, which is shifting requirements and introducing consideration of disaggregated space architectures. The drive for resilience adds pressure and complexity to space acquisition.

Translating resilience into requirements for space programs may be challenging. Resilience is a feature of the overall space enterprise (space, launch, ground, and user segments). Achieving overall enterprise resilience requires rigorous systems engineering and explicit tradeoffs. There is no uniform definition of resilience or of how much is

---

[29] AFSPC, 2013.

needed. Thus, translating resilience into the system specifications that drive technology development, EMD, and production may create problems in requirements flow-down as observed in SBIRS.[30] An architecture based on diverse satellites (and diverse ways of getting satellites into orbit, such as hosted payloads) may lead to complex integration issues.

As discussed in Chapter Three, austere budgets tend to reduce acquisition and technical expertise. They also tempt the reintroduction of themes from the 1990s' acquisition reform era that contributed to significant difficulties in the programs, e.g., leveraging commercial systems and standards, and streamlining acquisition processes. We note, though, that it was not the *principles* of adopting commercial norms that led to problems, it was their implementation, and the risk of these problems recurring now that the major DoD space programs are in their production phase is muted. For development programs, though, the application of these principles requires reconsideration based on the degree of risk and the tolerance for failure associated with any particular program. Existing DoD space acquisition policies and space system building standards, practices, and procedures may not necessarily accommodate an acquisition model that includes commercial-like systems, systems with shorter mission life, and hosted payloads. A different model may have to be developed and prototyped to avoid misapplication without understanding the consequences, as was the case in the 1990s. Comprehensive analyses may be needed on how best to integrate the two models, or how to phase from the old model to the new one.

Over the near and medium term, what DoD launches into space tomorrow will look much like what it launches into space today. Because no new programs of record are planned any time soon, those with skills in technology development and EMD may leave the industry, and those now working in production may depart preemptively. In the end, the systems engineering and test population are at risk of becoming hollowed out, as they were in the 1990s. The contractors we interviewed argued that lack of investment in development could affect the "readiness" of the industrial base to develop next-generation space systems. The recent Department of Commerce space industrial base survey revealed that many suppliers are not investing in research and development, thus affecting their design expertise.[31] As a result, the inevitable need to rebuild today's space constellations may be extended and expensive. But our interviews with acquisition officials show that there are no funds to pay for a low-level program of record designed to retain the talents that otherwise might be lost.

---

[30] One definition of resilience is offered in AFPSC, 2013.

[31] U.S. Department of Commerce, 2013.

Summary

In recent years, progress has been made in space acquisition. The five programs we examined are transitioning into the production phase and satellites are being launched.[32] In part for this reason and in part because of recent efficiency and cost-saving initiatives, costs and schedules are under better control, but they are still considered unaffordable by acquisition authorities. Moreover, emerging threats and severe budget cuts may create their own challenges. The Air Force is considering a gradual move to a disaggregated architecture with smaller satellites and commercially hosted payloads to cope with these challenges. The net benefit of a disaggregated architecture is yet to be determined, as noted by senior leadership.[33] Some characteristics of disaggregated architectures could alleviate today's risk and complexity factors in space acquisition but not without creating new risk and complexity factors (such as increasing complexity in synchronization and constellation management because of diversity in the architecture). Many challenges that future space acquisition faces are similar to those faced in the 1990s, which created a strained acquisition environment that fostered a high-risk acquisition approach. As such, future space acquisition is exposed to the risk of following the same path if the lessons learned from past acquisition experience are not adequately applied.

---

[32] As of fall 2013, GPS III-1 is in the final stages of development integration and testing and GPS III-2 is at the beginning of the production line (SMC, 2013b).

[33] AFSPC, 2013.

CHAPTER SIX

Conclusions

Military satellite systems have an irreducible complexity that make them susceptible to cost overruns and schedule slips. Space is a tough environment—it is a vacuum that lacks the protection against radiation and temperature fluctuations that atmosphere provides. Getting into space involves significant challenges and costs. If satellites fail to reach orbit, there is no opportunity for a second attempt. If satellite hardware fails or falters in orbit, it is nearly impossible to repair cost-effectively; thus, the satellite has to be either perfect or highly redundant when launched. The high launch cost also fosters aggregation of multiple payloads and missions onto a single platform and long mission life to maximize the satellite's capability in conditions of minimum space and weight, further increasing the complexity of the space system. Finally, making space systems work is more than building satellites; their delivery schedule must be coordinated with the availability of launch services and the delivery of ground equipment for both command and control and user reception.

In the 1960s and 1970s (and 1980s, to some extent), the field was new and the fielding of space capabilities was exploratory. The Cold War made it so important to get satellite capability into orbit that DoD could tolerate some launch and on-orbit failures in the hopes of getting capability in place as soon as possible. The end of the Cold War and experience gained in the first Gulf War made life more complex for the space acquisition community. It had more customers with more demands. This meant more satellite features and thus a far more complex systems integration problem. However, DoD had less money to satisfy the burgeoning demand. DoD tried to meet these conflicting needs by emphasizing more commercial elements in its acquisition strategy, e.g., by substituting insight into the contractor's performance for oversight of the details of that performance. But that approach introduced high risks into the program, which were realized when problems were caught later in the EMD cycle, which led to costly and time-consuming rework. The post–Cold War ebbing of technical capability on both sides of the government-contractor line further exacerbated the problem.

After about a decade of that approach, DoD shifted "back to basics," and satellite systems acquisition appeared to have stabilized. However, the stability in the satellite systems that were troubled early in their development—SBIRS, AEHF, GPS IIF, and WGS—may be partly due to ending development and entering production.

The current era of program stability may be short-lived, and optimism may be premature. Emerging threats are again shifting requirements and architectures for the space enterprise, driven by the need for increased resilience. The drive for resilience is adding pressure and complexity to space acquisition. Further, as in the 1990s, austere budgets are likely to reduce acquisition and technical expertise. They also tempt the reintroduction of themes (e.g., leveraging commercial systems and standards and streamlining acquisition processes) from the 1990s' acquisition reform era that contributed to significant difficulties in the programs. Several interviewees noted, though, that it was not the *principles* in the acquisition reform that led to problems, it was their implementation.

For development of future programs, the application of any acquisition efficiency initiatives requires a careful assessment based on the degree of risk and the tolerance for failure associated with any particular program. Moreover, although some characteristics of disaggregated architectures could alleviate today's risk and complexity factors in space acquisition, other new risk and complexity factors are introduced (such as increasing complexity in synchronization and constellation management as a result of diversity in the architecture). The overarching conclusion is that there is no "silver bullet" to fix space acquisition difficulties. All realistic acquisition approaches require tradeoffs and the assumption of some risks. Comprehensive analyses to inform such tradeoffs (including tradeoffs at the enterprise level) are needed for a robust acquisition approach.

List of Interviews

Table A.1
Interviews Conducted in Support of the Research

Organization	Topic
Space and Missile Systems Center	Enterprise management
Space and Missile Systems Center, Space Development	Enterprise management
Aerospace Corporation	Enterprise systems engineering, general space acquisition issues
Office of Secretary of Defense Space and Intelligence Office	General space acquisition issues
Office of the Assistant Secretary of the Air Force for Acquisition	General space acquisition issues
Office of the Deputy Under Secretary of the Air Force (Space)	General space acquisition issues
Office of the Secretary of Defense Cost Assessment and Program Evaluation	General space acquisition issues
Lockheed Martin Space Systems Company	Industry view
Boeing Space and Intelligence Systems	Industry view
Northrop Grumman Corporation	Industry view
Joint Staff Force Structure, Resource, and Assessment Directorate	Requirements
Headquarters Air Force Operations, Plans, and Requirements	Requirements
Headquarters, Air Force Space Command	Requirements
McKinsey & Company	Space acquisition cost growth study

Projecting Defense Space System Budget Growth: Issues of Inflation Index Selection

Background

Definition of "Real" Space Systems Cost Growth

In Chapter Two, we explain the systemic as well as unique system acquisition root causes for cost growth and schedule slips from the contractor teams' ATP through development, ground environmental testing, production, and launching of the first block of military satellites. For the majority of Air Force space system programs we evaluated, the primary root causes for "real" cost growth were government scope changes ranging from requirements "creep" to quantity increases, contractor challenges in mitigating a variety of technical risks, or some combination of the two.

In all cases, we measured "real" space system acquisition cost growth by comparing the latest SAR government program manager's estimated contractors' total prices at completion with the total of the initial contract prices awarded at start of the EMD plus the initial and follow-on production contracts in the previous SARs listed in then-year (TY) dollars and then converting those prices to inflation-corrected constant base-year (BY) FY 2013 dollars.

We applied the same inflation index used in the SARs for all MDAPs, which is reflected as the differences between the program's BY[1] and TY RDT&E and Procurement budgets and in the PAUC and APUC listed in both BY and TY dollars.

DoD Comptroller's Derivation of Appropriation-Specific Inflation Indexes

The acquisition data reported in each SAR use the same Bureau of Economic Analysis (BEA) Gross Domestic Product (GDP) price index values[2] that the Office of the Under Secretary of Defense Comptroller (OUSD [C]) has been using as the primary

---

[1] The program BY budget data reported in the SARs are usually set as the same fiscal year as approval of the MDAP major milestone for go-ahead into the EMD or production phase is approved, in most cases concurrent with the initiation or update of the APB.

[2] The GDP price index is based on the market basket of the sum of U.S. economic changes over time of consumption, investments, government spending, and exports minus imports of all domestic goods (products) and services provided to final users. The index is quantitatively defined as an economic annual set of metrics or measures that accounts for inflation by converting the prices of all new, domestically produced goods and services in

basis for deriving appropriation budget-specific price indexes for the national defense budget submission to Congress.[3] As part of the national defense budget estimates, the appropriation-specific price indexes are reported by the OUSD (C) and updated each fiscal year as DoD deflators covering seven years: the two prior years, the current budget year, and four out-years for RDT&E and Procurement budgets.[4]

Importance of Selecting Relevant Inflation Index

Appropriation-level DoD deflators are issued to the military services and DoD agencies in an annual OUSD (C) "Revised Inflation Guidance" memorandum for putting together the programming phase of the military services' RDT&E and Procurement annual Future Years Defense Program President's Budget requests for each program element in annual TY dollars. The military services' assistant secretary for financial management or equivalent office usually develops appropriation-specific inflation indexes based on DoD deflators and passes them on for use by the major acquisition commands.[5]

In the Air Force, the Under Secretary of the Air Force for Cost and Economics (SAF/FMC) develops the Air Force–wide raw inflation indexes by appropriation category (e.g., RDT&E [3600] and Aircraft and Missile Procurement [primarily 3020 for space systems]) based on the OSD inflation guidance set of corresponding raw index values.[6] SAF/FMC requires that all the acquisition commands for major weapon system procurements base estimated prices in annual TY dollars using the SAF/FMC inflation indexes.

---

the U.S. economy into constant dollar prices. The GDP deflator index is published quarterly (see BEA, undated (c), for the latest quarterly Table 1.1.4 update of "Price Indexes for Gross Domestic Product."

[3] Horowitz et al., 2013. The Office of Management and Budget (OMB), the Council of Economic Advisors, and the Department of the Treasury provide joint annual guidance to OUSD (C) on the recommended use of the projected values of the GDP price index, based on composite rates accounting for nonpay factors for deriving DoD-wide appropriation-specific price indexes as the recommended economic measure for forecasting annual inflation rates of change on purchase prices of all defense weapon system products.

[4] See DoD, 2012. Chapter 5 documents the OUSD (C) office's treatment of inflation by providing DoD deflator tables to be used as inflation indexes for converting the requested DoD weapon system budgets from constant year to TY dollars over the Future Years Defense Program. Specifically, Tables 5-5 and 5-7 in Chapter 5 respectively list two columns of annual DoD deflator values for procurement and RDT&E total obligation authority and budget authority as standard economic factors for calculating annual inflation rate changes for defense product (or system) acquisition programs. For example, the deflator value for BY FY 2013 is set at 100 and includes a listing of lower annual values going back to FY 1970 and higher annual values through FY 2017. These two columns of deflator values are for DoD development and procurement of all defense weapon system products and do not include any annual economic changes in government pay, fuel, and medical costs. The budget authority reflects the annual budget appropriated by Congress to each program, and the total obligation authority reflects the available annual budget that each program has available to spend within a fiscal year.

[5] Horowitz et al., 2013.

[6] AFI 65-502, 1994.

Any exemptions from using the SAF/FMC inflation indexes would require that the System Program Office (SPO) justify why OSD-derived inflation rates should not apply to its program and why its proposed system-specific rates should apply. The SPO would also have to provide its sources and methodology and gain the concurrence of the acquisition command's financial management offices. Once approved by the SAF/FMC office, the SPO request would go to OSD to approve the exemption. We discuss the details of the current inflation guidance policy and practices with the Air Force's Space and Missile Command below, in the section on inflation policy guidance.

OSD Cost Assessment and Program Evaluation (CAPE) office's cost community also uses DoD deflators in generating constant and TY Independent Cost Estimates (ICEs) for MDAPs for major milestone reviews and when Nunn-McCurdy cost breaches are issued. An ICE is required as part of the program recertification process. CAPE is also required to "periodically assess and update the cost (or inflation) indexes used by the Department to ensure that such indexes have a sound basis and meet the Department's needs for realistic cost estimation."[7] We discuss its recent assessment of inflation practices across the military services below, in the section on current military services' inflation practices.

For a space system program to be fully funded, or any DoD weapon system program for that matter, adequate funding must be appropriated up front to cover all projected future TY costs for the portion of the program authorized in a given year for acquiring a specified quantity of space satellite and ground segment deliverables. If acquisition command government budget planners use the DoD deflators as the basis for generating TY program budget requests, they could misestimate the potential fluctuations in authorized program funding over time, because of unanticipated higher (or lower) rates of defense-wide inflation in defense sector–specific contractors' changing labor rates and higher (or lower) material estimates compared with their negotiated prices. If annual program funding appropriations fall short, the program could overrun its total budget. If this pattern of contractor expenditures outpacing program funding consistently occurs for acquisition of products across, or specifically within, a given defense sector, the current OUSD (C) inflation guidance policy and CAPE practices of using DoD deflators as a basis for measuring a program's inflation-adjusted "real" cost growth could come into question.

How the Appendix Is Organized

The next section of this appendix lays out our specific space cost inflation research objectives, provides a focused set of research questions, and describes the research approach in answering these questions. The next two sections provide responses to

---

7   Public Law 111-23, 2009.

our research questions on the rationale or motivation and the economic basis for using and setting defense space-unique inflation indexes. Two more sections follow, giving responses to the remaining two questions on the relevance of using space inflation indexes rather than defense-wide indexes based on DoD annual deflators, and progress on validating space inflation indexes as a more accurate basis for measuring a program's "real" cost growth. We conclude the appendix by providing our overall findings and suggested actions for ensuring that projected annual inflation rates are relevant and accurate predictors for estimating the future acquisition cost of defense space systems.

Research Objectives and Approach

The objective of our research is to provide OSD (AT&L) PARCA office management with a more informed understanding of the Air Force Space and Missile Systems Center (SMC) and NRO financial management and cost communities' current rationale for setting defense space inflation indexes and their processes for ensuring that these indexes are accurate for converting constant year system cost estimates to projected TY dollars for their ongoing and future space systems' acquisitions.

The research team's approach, described below, was focused on gathering information from key government space cost community experts in the budget planning and estimating process by soliciting responses to the following questions:

- *Rationale for using defense space inflation indexes:* Why has it been necessary to use defense space-unique inflation indexes rather than GDP price index–based DoD deflators or other defense sector price indexes?
- *Economic basis for setting defense space inflation indexes:* What is the economic basis and approach taken for setting defense space-unique inflation indexes?
- *Relevance of defense space inflation index for measuring real cost growth:* Are current defense space inflation indexes accurate in converting TY, ongoing, or estimated costs to constant dollars for measuring a space system program's "real" cost growth?
- *Validating defense space inflation indexes:* Have the space inflation index projections been validated in a comparable way to GDP price index–based DoD deflators?

Our research approach to answering these four questions had two components. The first consisted of reviewing recent documents and briefings covering current government financial management practices that have been considered or used across the military services in setting program budgets and estimated TY dollar prices of products across different defense sectors. This literature review provided information on both the economic factors and discriminating business base differences that have been used within the different defense industrial base sectors to justify using product-unique inflation indexes rather than the DoD deflators as the basis for setting inflation rates.

The second component was a series of "not for attribution" structured interviews of government experts in the defense space cost community. Before meeting with the defense space cost government organizations listed in Table B.1, we sent out a read-ahead set of inflation-related questions focused on gathering information and responses to the four research questions listed above.

As part of the interviews, we used a schematic summary of our views, illustrated in Figure B.1, to elicit tailored discussions on a rational process for deciding to use a defense space cost inflation–unique index rather than the GDP-based DoD-wide deflators or other defense or producer price indexes and to justify why this different index better correlates with historical cost growth. Further, we used this schematic to point out and get feedback on our views on the need for finding a disinterested, vali-

Table B.1
Government Organizations Visited

U.S. Bureau of Labor Statistics, Division of Price and Index Number Research

OSD CAPE office

Air Force Cost Analysis Agency, Space Technical Experts

Air Force SMC/FMC office

Office of Director of National Intelligence (ODNI) Cost Analysis (CA), Directorate office

NRO Cost Analysis and Assessment Group (CAAG)

Figure B.1
Space Systems Cost Index Logic

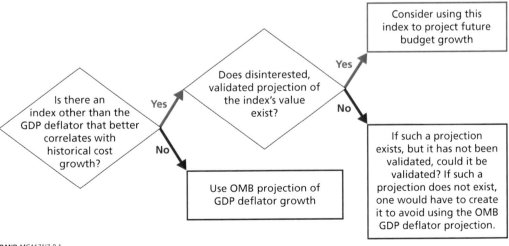

dated forecast of the chosen index.[8] Such a forecast must be validated, i.e., the chosen index must have been shown to result in an unbiased set of estimates of index values based on historical data.

The remainder of this appendix provides answers to the four questions posed above. Following those sections, we present our findings.

Rationale for Using Defense Space Inflation Indexes

This section explores the reasons for using space inflation indexes rather than the GDP deflators. It begins with a discussion of why DoD deflators should not be used, then compares alternative defense indexes, makes some historical comparisons, and lays out the rational for using space indexes illustrating the rationale with one program, WGS.

Opposition to Using DoD Deflators

Several arguments have surfaced over the years that question the relevance of the GDP price index as the sole basis for projecting DoD deflator values into the budget out-years.

One argument holds that as a defense program progresses, its costs might be expected to change, even if the functions being performed or goods being produced are static, reflecting stable customer/user requirements. On the one hand, costs could decline. For example, production labor efficiency learning curve effects could make the incremental costs of the 1,000th unit lower, in real terms, than the incremental cost of the 100th unit. GDP and other government price indexes should take into account DoD procurement of products where the range of quantities and effects of the production learning curve efficiency can vary significantly across defense sectors and can drive unit prices.

Another argument says that the effects of prices on military-unique products differ from the price effects of buying commercial-off-the-shelf products, where prices are primarily a function of market-driven supply and demand for goods and services, such as personal computers. Personal computers are an example of a price-declining product where, controlling for quality and capability, prices have consistently and markedly declined. DoD would not be viewed favorably if it were paying the same amount for a personal computer of the same capability today as it paid in 2000.[9] As we discuss in the next section, commercial products, such as personal computers, and analogous

---

[8] The term *disinterested* is used in this context as an example of selecting a group of government economic analysts, such as those in the Congressional Budget Office (CBO), who have no vested interest in the specific financial outcome of any defense space systems acquisition. The group can provide an unbiased and impartial professional assessment for validating, by consensus, that the results of the projected set of space inflation index values are accurate.

[9] Worthen, 2010, suggests that, in recent years, consumers have been more likely to experience increases in the capabilities of personal computers than actual declines in their nominal prices.

military systems across different industries and product segments have vastly varying inflation rates, as signified by the very extensive and fine-grained set of Producer Price Index (PPI) series.

The ability to substitute by purchasing a different product does mitigate the effects of incurring market-driven price fluctuations in the broader commercial product sector of the economy. However, as a previous RAND study pointed out, defense operators have fewer opportunities to switch to different providers and products when prices increase.[10] Therefore, the authors concluded that DoD systems can experience higher inflation than the general economy, leading to higher costs.

Finally, as we have seen for space system programs, some prices over the course of the acquisition phases have increased faster than overall inflation. If an annual DoD contract involves such an "above average cost growth" item, it might be reasonable to adjust its associated annual budget upward to account for this supernormal cost growth. As we pointed out in Chapter Two, the SBIRS program and the GEO satellite and HEO payload deliverables have experienced above-average cost growth, clearly outpacing the average annual GDP price index values or any other sector-unique government alternative price indexes, which we discuss next.

Retrospective Comparisons for Using Alternative Defense Sector Price Indexes

At present, the Bureau of Labor Statistics (BLS) tabulates ex post values for a variety of PPI series for military and analogous commercial systems. However, relatively few other government-provided price index projections besides the GDP price index cover only a small number of aggregated series. For example, the CBO provides annual baseline projections of the Personal Consumption Expenditure Price Index, Consumer Price Index, and the Employment Cost Private Wages and Salaries Index.[11]

In recent years, different defense product sectors have used different BLS PPIs with values that are higher or lower than the projected GDP price index multiplier or deflator values.[12] In addition to the GDP price deflator, the BEA publishes deflators for procurement of five major types of military systems: aircraft, missiles, ships, vehicles, and electronics.

---

[10] Connor and Dryden, 2013. In addition to the general conclusions noted above, the RAND report specifically observed that U.S. Army Bradley parts costs have risen faster than operating and support official budget inflation, whereas U.S. Army Abrams parts costs have risen slightly less quickly than operating and support official budget inflation.

[11] See the CBO quarterly updated baseline index projections as downloadable datasets (CBO, 2014).

[12] *Deflators* for GDP and other government producer price and other indexes are a normalized set of values similar to the annual published set of DoD deflators. For DoD deflators, a normalized value of 100 is set at the current budget base year with all the previous fiscal years lower and projected out-years higher than 100 to reflect, respectively, relative previously lower and higher annual inflation changes over the base year. For retrospective comparisons of annual projections across annual indexes, deflator values are all set to a normalized value of 100 for the first year to provide relative comparisons of increases or decreases in index values over the same span of fiscal years.

Table B.2 compares the GDP deflator values representing the annual rate of inflation with several BEA defense indexes and the BLS PPI trajectories most relevant to different types of defense products during the 1985 to 2009 25-year time frame.[13]

On the one hand, the BEA indexes for defense aircraft, defense missiles, and defense electronics since 1985 have each grown much less and at slower rates than the GDP deflator index. However, research by the Institute of Defense Analyses noted that these three indexes are highly suspect, since annual values depend on how system costs associated with improvements in capability of the defense product's value over time are measured and then used to normalize the annual differences in the rate of growth of the prices for each category.[14] On the other hand, the BLS PPI trajectories for aerospace products and parts and defense ships are quite close to the GDP index deflator annual rate of inflation.

We make two observations. First, the policy implication of these comparisons is that the differences in annual inflation growth rates among the defense and defense-related indexes suggest that DoD might obtain better measures of the "real" value of

Table B.2
Defense and Producer Price Index Deflators Related to Defense

Deflator	Average Annual Growth Rate, 1985 to 2009 (%)	Total Growth, 1985 to 2009 (%)
GDP	2.4	78
BLS PPI for Defense ships	2.7	90
BLS PPI for Defense vehicles	1.9	56
BEA Defense aircraft[a]	0.1	1
BEA Defense missiles	−0.3	−8
BEA Defense electronics	−1.5	−31

SOURCES: The data in the table are from 360 sector databases developed by Inforum (undated). The DoD values are from Inforum's "Federal Defense" table and the economy-wide figures are from Inforum's "National" table. The "National" table combines spending for federal defense, federal nondefense, nonfederal government, and the private sector.

[a] The BLS does not publish indexes for military aircraft, because there are not enough domestic producers to meet the BLS's standards for survey respondent confidentiality and statistical accuracy of the index.

---

[13] Horowitz et al., 2013. The values listed in Table B.2 are extracted directly from Table 2, and the reference cited for the source of the data is listed in table note (1) and footnote 93.

[14] Since there is wide variability, we believe that there is a need for further research to gain more insight into the details of how the normalization is done in specific defense sectors by adjusting over time based on a product's capability and value to the defense customer.

cost growth across MDAP weapon system budgets by using sector-specific alternative price indexes instead of the GDP price index–based DoD deflator.[15] Second, as we discuss further in the next section, the comparison of price index deflator values does not cover the specific portion of the defense sector that deals with military and intelligence community space systems products.

Historical Comparisons

Beginning as far back as the mid-1990s, there has been an ongoing concern within the government space cost community that actual space system inflation experiences do not always reconcile well with computed OUSD (C) annual inflation rates based on RDT&E and procurement appropriation budget–specific DoD deflator values (excluding pay, fuel, and medical). Comparison of inflation rates showed that space inflation rates exceeded rates in other sectors. From 1992 through 2000, the average annual RDT&E (AF 3600[16]) inflation rate was approximately 1.8 percent compared with the NRO CAAG office's computed average annual space inflation index of 3.0 percent (1.2 percent higher) over this same 13-year time frame.[17] From 1992 through 2000, the average annual space inflation index was slightly lower, at 2.9 percent, but still 1.1 percent higher than the average RDT&E (AF 3600) inflation rate. However, from 2001 through 2005, the average annual space inflation index rose to 3.2 percent, which is 1.4 percent higher than the RDT&E (AF 3600) average rate of 1.7 percent over this same time frame.

Rationale and Motivations

Careful data analysis indicates that the NRO experienced higher inflation over this 13-year time frame (1992 through 2005) because of the unique characteristics of the defense space systems acquisition and industrial base workforce.[18] These characteristics include the following:

- Development primarily uses "state-of-the-art" (and beyond) technologies.
- Only limited production quantities are required for manufacturing, and in most cases they are customized, "one-off" products lacking the benefits of mass production learning efficiencies and productivity effects.
- Space products with military and intelligence community mission-unique payloads have very limited commercial market potential, which means that the

---

[15] Horowitz et al., 2013.

[16] AF 3600 denotes specific Air Force investment appropriation funds covering RDT&E, which is different, for example, from funds for Air Force aircraft procurement using AF 3020 appropriate funds.

[17] Hogan, 2013.

[18] NRO, 2009.

majority of defense space system contractors cannot spread overhead costs over a broader business base.

- Secure manufacturing facilities and trusted source components are required.
- Clearances and the associated security costs to retain critical skilled personnel are needed.

In more recent discussions, other government space cost experts have confirmed some of the NRO experiences and pointed to other reasons that drive up space inflation indexes:

- Contracts for acquiring deliverable quantities of defense space systems are smaller (in single digits) than typical low- to full-rate production quantities for military aircraft, missiles, and other defense sector products.
- Space program annual budget overruns could be the result of increases in the contractor teams' costs as well as the actual rate of expenditure of the direct labor and escalated material costs, because the period of performance can be longer than originally planned. Additional time can be required for designing defense space-qualified hardware, developing mission-unique software, performing ground environment subsystem and system satellite testing, and delivering the systems to launch sites in preparation for launches.
- Furthermore, after the initial satellite block build funded by the NRO and in most cases by the Air Force with RDT&E budgets, there could be unexpected production delays and increased startup costs in the next block build of satellites, a slowing of build rates, and supplier parts obsolescence delays, all of which may also increase program costs.

The WGS program provides an example of the representative effects that basing program budget TY dollar inflation rates on defense space sector inflation-based fluctuations in contractors' labor or material price expenditures over the contract period can have on measuring the magnitude of "real" constant-year-dollar cost growth compared with using inflation rates based on DoD deflators.

Wideband Global SATCOM Program Cost Inflation Example

For the WGS program, the space system contracts from Block I forward were all awarded as firm fixed price contracts where the contractor, Boeing, had and continues to have the responsibility of managing the technical, ground testing, and technical risks of controlling costs and meeting launch window schedules. The WGS average unit price for the Block II follow-on satellites increased beyond what would be expected using DoD deflators, even though they were essentially the same as the previous Block II versions.

A previous RAND OSD (AT&L) PARCA office–sponsored study provided a detailed explanation of the primary reasons for the unit cost increase between the

Block II follow-on satellites seven through ten over the previous unit cost of the two previous Block II satellites.[19] The Air Force SMC WGS program office provided the RAND study team with unit cost estimates, one for Block II satellites at a target price of $355 million in BY 2007 dollars and a BY 2011 Block II follow-on unit ceiling price of $420 million.

Before providing a quantitative breakdown to account for the differences in the higher unit cost of Block II follow-on satellites over Block II, the Block II unit cost in BY 2007 first increased from $355 million to $366 million to account for a 3 percent reported overrun above the target price. Table B.3 lists the adjusted Block II satellite unit cost estimate of $366 million in the first row along with three different Block II unit cost estimates listed in the rows below that have all been applied to convert this BY 2007 estimate to the same BY 2011 dollars using different average annual inflation rates all compounded over the same four-year time frame

The RAND study team used an average inflation rate of 3.5 percent a year over a four-year period (from FY 2007 through FY 2011) to convert the expected unit cost of Block II satellites in BY 2011 dollars to $420 million. The WGS program office supplied the study team with this inflation factor value, which was based on Boeing's historical experience in satellite component and manufacturing costs over the Block II

Table B.3
Inflation-Based Differences in WGS Block II Satellite Expected Unit Prices

	Average Annual Rate (%)	Inflation Factor Covering 2007–2010	WGS Block II Satellite Unit Price ($ millions)[a]	
Actual unit cost, BY 2007, $ millions[a]			366	Increased "Real" Cost Growth over PM's Inflation-Based Unit Price (%)
WGS PM Boeing inflation value	3.5	1.147[b]	BY 2011	
WGS PM expected unit cost,			420	
DoD deflator circa 2011	1.8	1.074[c]	BY 2011	
OSD-based unit cost			393	6.8
DoD deflator, FY 2015 budget, April 2014	1.6	1.065[d]	BY 2011	
Revised OSD-based unit cost			390	7.7

[a] Representative WGS FY 2011 estimates and inflation index values were extracted from Table 6.5 in Blickstein et al., 2001, for the "Actual unit cost, BY 2007 $" estimate of $366 million in FY 2007 BY dollars and the "Expected unit cost circa 2011" (Blickstein et al., 2001).
[b] 1,035 x 1.035 x 1.035 x 1.035 = 1.147 x $366 million = $422 million.
[c] 1.018 x 1.018 x 1.018 x 1.018 = 1.074 x $366 million = $393 million.
[d] 1.016 x 1.016 x 1.016 x 1.016 = 1.065 x $366 million = $390 million.

---

[19] Blickstein et al., 2011.

production period from 2007 through 2010. The PM-supplied inflation factor value exceeded a 1.8 percent average annual inflation rate based on DoD deflator values at the time of the RAND 2011 study for converting FY 2007 constant dollars into BY 2011 dollars over that same period. Using the lower deflator value resulted in an expected Block II satellite unit cost in BY 2011 dollars of $393 million. In real cost growth terms, the WGS PM's Block II unit cost estimate was 6.8 percent higher than the same FY 2011 constant dollar estimate using DoD deflators.

If one does the same constant dollar conversion from BY 2007 to FY 2011 dollars using the current FY 2015 DoD deflator-based average annual inflation rate, the rate declined from 1.8 percent to 1.6 percent over the same four-year period yielding a revised cost of $390 million.[20] In real terms and based on the WGS PM's use of Boeing's inflation values over the most current DoD deflators, the Block II satellite unit cost growth increased by another 0.9 percent from 6.8 percent to 7.7 percent.

In the next section, we describe the current basis that the defense space cost community is using for setting space inflation indexes.

Current Economic Basis for Setting Defense Space Inflation Indexes

This section explores the economic basis for defense space inflation indexes. It begins by using OSD and Air Force guidance for such indexes and then describes the NRO and SMC practices in setting such indexes.

OSD and Air Force Guidance for Setting Defense Space Inflation Indexes

In FY 2010, the CAPE office commissioned an independent study to assess the treatment of inflation across the military services and SPOs managing MDAP acquisition costs.[21] The CAPE office pointed out that some DoD organizations, most notably Naval Sea Systems Command, Naval Air Systems Command, and many Air Force program offices, have developed specialized inflation projections for their programs. The study also noted that the SPO projections are usually higher than the approved DoD deflators. The CAPE study observed that if these programs experienced inflation in line with the program-specific projections but higher than the DoD deflators, then the programs were systematically underfunded, leading to high "real" program cost growth.

Their study findings, on the one hand, stated that with oversight from the CAPE and the OUSD (C) offices, it would be appropriate for the SPOs to use program-specific inflation projections for managing their program baseline cost estimates and preventing them from growing. On the other hand, the CAPE also found it appropriate to use DoD deflators to calculate and report MDAP budgets in constant BY dollars

---

[20] OUSD (C), 2014.

[21] CAPE, 2012.

and to estimate program budgetary cost growth in assessing Nunn-McCurdy unit cost thresholds. CAPE stated that the approach to using DoD deflators conforms to OMB's guidance of reflecting DoD product prices in constant-year dollars to reflect the general purchasing power relative to the U.S. economy as a whole.

As mentioned above, the SAF/FMC policy guidance for inflation permits the SPOs to request from OSD an exemption from using DoD deflators for their proposed program-specific inflation index.[22] Similar to Naval Sea Systems Command for projecting Navy shipbuilding costs and Naval Air Systems Command for estimating naval aircraft budgets, SMC has recently adopted its own inflation projections for estimating time-phased annual program acquisition budgets in TY dollars and in translating TY budget submission requests back to constant BY dollars.[23]

In November 2011, the commander of SMC provided a guidance memorandum[24] that references the SAF/FMC 1994 inflation policy guidance instruction and permits program offices to use different approaches, which included using inflation rates authored by NRO, the CAPE offices, and the Defense Contract Audit Agency (space industry plant-wide forward pricing rate proposals, agreements [FPRAs],[25] etc.) when developing their space program TY cost estimates. The SMC cost guidance went on to state that until further guidance or refined approaches to estimating space cost inflation rates are developed, the program offices should use the most applicable rates. In their documentation asking OSD to grant exemptions to the use of Air Force OSD-derived inflation rates, the program offices must explain the rationale for applying alternative rates to their program and the cost effects of using these rates. The SMC guidance goes on to state that when program offices develop TY estimates using their proposed inflation index, they must still use Air Force OSD-derived inflation rates to convert TY estimates to program BY dollars.

NRO and SMC Practices for Setting Space Inflation Indexes

As pointed out in a previous report,[26] the NRO CAAG office in 2002 selected a firm outside the government, Global Insight,[27] to provide annual cost index projections as the source data basis for computing their agency's space cost inflation rates.

---

[22] AFI 65-502, 1994.

[23] Horowitz et al., 2013.

[24] Pawlikowski, 2011.

[25] An FRPA is a written agreement between contractors and the government—usually the Defense Contract Management Agency (DCMA)—to use certain rates during a specified period, typically two to three years for pricing future contracts and contract modifications. For further details, see DCMA, undated.

[26] Horowitz at al., 2013.

[27] Global Insight is a private firm that performs economic and financial analysis and forecasting. NRO CAAG and the Air Force SMC/FMC offices base their space systems inflation rates on Global Insight's ten-year predictions of direct labor and material price indexes.

During our interviews with the NRO CAAG office, they described their use of four Global Insight labor indexes and several commodity or material indexes, such as electronic parts. Global Insight updates indexes semiannually and as warranted based on ten-year rolling forecasts of the labor and commodity values. The NRO CAAG office annually updates its weighted labor, material, and other direct charges composite set of annual space inflation values based on the latest set of Global Insight updates.

The Office of the Director of National Intelligence/Cost Analysis, the Air Force Cost Analysis Agency, Air Force SMC/FMC cost personnel conveyed during our interviews that they were all using Global Insight annual index projections for covering a variety of disaggregated inflation index projected values for direct labor, direct material, and other direct charges (ODC). Similar to the NRO CAAG office, these government space community offices are using their space inflation index for the following purposes:

- normalizing historical space program as-spent costs to constant year dollars within their space system parametric cost model databases
- time-phasing ODNI/CA ICE constant year cost results and cost position estimates from Air Force space program offices and AFCAA to annual, TY dollars.

Defense Space Inflation Space Index Relevance for Measuring "Real" Cost Growth

This section assesses the relevance of space inflation indexes for measuring cost growth. It begins by describing how OSD validates the GDP price index and then discusses the validation of defense space inflation indexes.

OMB Process of Validating GDP Price Index

In our discussions with the government space cost community, we described the OMB validation of the GDP price index illustrated below as a representative example for doing a similar periodic validation of their space cost inflation indexes.

OMB projects the GDP deflator index value for five years and characterizes those forecasts as being "fairly accurate" without any marked upward or downward bias.[28] In 2010, OUSD (C) performed a prospective, analytically based assessment of the accuracy of the initial set of estimates of GDP price index. The results are displayed in Figure B.2 in terms of the GDP annual growth rate values from 1991 through 2009. The GDP annual growth rate values forecasted five years in advance (noted by the dashed blue line) are compared to current-year actual computed rate estimates (as of

---

28   Horowitz et al., 2013.

2010) (noted by the solid blue line) with the values of the annual differences between the two estimates plotted as the solid red line over the same time frame.

The figure shows the accuracy of these GDP growth rate forecasts over the past 19 years. The three lines represent the trend of annual GDP growth rate differences between the initial set of forecasted values predicted five years in advance (displayed by the dashed blue line), the actual computed growth rate as of 2010 (displayed by the solid blue line), and the difference between the forecast and actual annual growth rates (displayed by the red line). For example, the initial forecast for the estimated GDP growth rate for 1991 was predicted five years prior at 2.3 percent (denoted as the value where the dashed blue line intersects the y-axis). The 2.3 percent forecasted 1991 GDP growth rate value compares with the actual 1991 growth rate higher value of 3.8 percent (denoted as the value where the solid blue line intersects with the y-axis). The difference between the predicted and actual 1991 GDP growth rate value is –1.5 percent (denoted as the value where the solid red line intersects with the y-axis).

Overall, the five-year rolling annual forecasts repeated over this time frame appear to be fairly accurate. The number of overestimates and underestimates were about the same (10 versus 9), and the absolute value of the yearly errors averaged only 0.8 percent.[29] The overestimates were a bit larger at a maximum value of 1.7 percent than the

Figure B.2
Validation Process Example of the Accuracy of GDP Inflation Rate Predictions

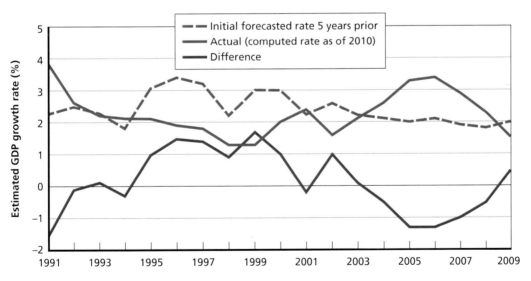

SOURCE: Horowitz et al., 2013.
RAND MG1171/7-B.2

---

[29] For clarification, the average yearly averages as a measure of accuracy or validation of the GDP annual growth rates are computed by adding up the cumulative negative and positive percentage value differences and dividing by the 19 years.

underestimates at 1.5 percent. Even though the estimates varied a good deal from year to year, they usually became more accurate as the year of execution approached.

Validating Defense Space Inflation Indexes

Even though the NRO CAAG supplied RAND with a previous 2009 briefing that documented its previous attempts to validate the accuracy of its composite space inflation index, we believe that without further details, which were not provided, it fell short of any detailed retrospective rolling assessment of past annual projected index values similar to the GDP annual growth validation example illustrated above.[30]

In a similar vein, the AFCAA more recently conducted a space cost inflation rate study as a follow-on effort to work done by NRO CAAG (supported by Global Insight, Inc.), which re-addressed the question of whether annual inflation rates in SMC space system programs were accurately reflected in OSD inflation indexes reported as RDT&E and Procurement DoD deflators. The results of the AFCAA study proposed using a composite defense space inflation index similar to NRO's based on using a weighted average of a comparable set of Global Insight labor, materials, and ODC index values.[31] The relative contributions, calculated using historical SMC space program data and best estimates, are as follows:

- 81 percent for labor (including labor overhead)
- 13 percent for material/purchased parts
- 6 percent for ODC.

Even though the AFCAA study in part endorsed and adopted an approach similar to that of NRO CAAG in computing a composite space inflation index, it also fell short of validating that this defense sector–unique index was more representative and relevant than GDP-based DoD deflators.

Findings

The five findings listed below summarize our research assessment of the collective responses provided by the government space cost community on this topic.

---

[30] NRO, 2009. NRO CAAG attempted to validate the need for a higher inflation index by justifying that the DoD deflator-based inflation index values were too low to correct for the time-variant historical annual residual differences between previous NRO cost-estimating relationship–based predictions over cost actual data points over a 40-year period from 1965 through 2005. Even though it provided an analytical argument for justifying the need for a space-unique inflation index, it fell short of validating the accuracy of its index.

[31] Hogan, 2013.

Ample Justification Exists for Using Space Inflation Indexes over DoD Deflators

As we discussed above, there is ample evidence within the government space cost community that defense and intelligence community customers acquiring space systems today are in a unique sector where the following conditions exist:

- The marketplace for acquiring products is limited to a few prime contractors and subcontractors.
- Only a few defense space companies have a workforce with the unique labor mix of highly skilled engineering and manufacturing personnel able to leverage the required state-of-the-art and beyond maturing technologies and with the security clearances needed for developing defense systems that have a unique set of space capabilities and mission-unique payload requirements.
- In addition, and depending on prime defense space systems providers, varying levels of commercial space market share limit a contractor's ability to spread overhead costs over a broader business base. In fact, defense space contractor overhead rates may be relatively higher than the rates of other defense contractors providing other products, since the overhead rates may cover the additional costs of maintaining secure manufacturing facilities.
- Defense space programs involve relatively higher labor costs and relatively lower material (nonlabor) costs.[32]
- The nonrecurring labor costs are proportionally more important. This phenomenon calls for using a labor-heavy (i.e., higher percentage labor-based) inflation index for estimating the defense space program economic rate of growth.[33] One government defense cost expert explained the need for more defense space contractor labor than commercial space contractor labor as representative of the added effort required to meet the higher level of screening and testing, additional quality assurance inspections, and increased documentation for acceptance of space-qualified parts, especially since there is less demand for these parts than for the more widely available commercial parts.
- The higher dependence on these types of space sector–unique, labor-intensive activities reduces contractor productivity gains seen elsewhere in other defense weapon systems technology sectors.[34]
- Finally, the space cost experts pointed out that space systems cost growth may also be presaged by defense contractors' use of FPRAs in a manner not observed

---

[32] Hogan, 2013. SMC, 2013. Space hardware contractors normalizing their own costs indicated that inflation rates could be reconstructed using a labor-to-material split of 65 percent to 35 percent.

[33] However, a labor-heavy inflation index does not axiomatically imply an inflation rate above the GDP deflator. For example, the BLS engineering services index (series 541330) has modestly lagged the GDP deflator since 2004.

[34] NRO, 2012.

in the overall economy. Even in the presence of an FPRA, there will still be future cost uncertainty, both as a result of nonlabor costs not covered by FPRAs and for periods of performance of space acquisition contracts extending beyond the typical two- or three-year term of negotiated FPRAs.[35]

Space Inflation Indexes Need to Be Validated Before Using Them to Measure Space Program "Real" Cost Growth

Given the findings above regarding the set of comparative differences, we find that identifying an alternative to using space system inflation indexes rather than DoD deflators for measuring "real" program cost growth is only a necessary, not a sufficient, condition.

In our view, an ideal space cost inflation index would have to achieve the following:

- reflect the higher labor cost proportion found in defense space system acquisition contracts
- use a disinterested annual forecast of the index's values that has been shown to be unbiased and acceptably accurate in retrospective analysis.

Given the current unmet need for space cost inflation indexes to be validated, our next finding assesses whether there is adequate evidence available to still rely on space inflation indexes as a more accurate, forward-looking approach to improving the projections for space systems acquisition cost estimates in TY dollars.

We Are Unable to Validate the Reliance of Space Inflation Indexes for Projecting TY Dollar System Costs over DoD Deflators

Even though there was no documentation on their validation process, NRO CAAG did provide details on the process it goes through every other year in ensuring that the current set of space contractors' labor and material weighting values for computing the composite set of annual inflation values accurately represent the projected space system industry's economic growth or changes (increases or decreases). It reviews contractors' labor-related basis of estimates, bills of materials, and FPRAs within cost proposals, along with data collected as part of the NRO-contractor Cost Integrated Product Team one-on-one sessions.

We were not given sufficient evidence or substantive data to be able to assess the accuracy and possible bias of the specific set of annual labor, material, and overhead cost projection indexes provided by Global Insight. At the time of this writing, we are not aware of any retrospective, unbiased, objective validation of Global Insight's projections by any of the space cost government offices we visited.

---

[35] Contractor FPRPs may provide insight beyond FPRAs, but those proposals for rates beyond the term of FPRAs do not represent the conclusion of a bilateral negotiation with the DCMA the way that FPRAs do.

We also note that, since Global Insight's methodologies are proprietary, tying government budgets to a private sector firm's "black box" forecasts is a concern. We believe that it would be preferable if cost projections were made by a disinterested government agency (such as the BLS or the BEA) using techniques that have been publicly vetted and reviewed.

Other Defense Sectors Face Similar Issues in the Use of Product-Unique Inflation Indexes

The arguments we have heard in favor of separately indexing defense space program costs are not unique to defense space systems. As a long-standing example of a sector-focused indexing, the Navy uses a Steel Vessel Index in its shipbuilding contracts exactly because of an observation that shipbuilding costs may evolve differently from the GDP price index-based DoD deflator. [36]The best inflation index for a defense space program will probably differ from the best inflation index for a shipbuilding program, but it is eminently plausible that an index other than DoD deflators may be appropriate in both contexts.

At least on a broad level, we do not think that there is anything extraordinary about defense space systems' need for a set of cost inflation factors different from the rates based on DoD deflators. Other parts of the DoD sector of different products might also have a mix of contractor, direct labor, and material cost that differs meaningfully from the broader acquisition appropriation-based DoD set of deflators.

We Are Concerned About Inflation Index Endogeneity: Is Budget Growth Self-Fulfilling?

One possible concern with defense space system program budget growth above the DoD deflators is that the growth will be self-fulfilling (or mutually reinforcing). A contractor who knows that its customer's budget will grow significantly higher than legacy systems might feel less pressure to contain its cost expenditures.

Unfortunately, the reverse experiment did not work. Experts we interviewed noted that actual space systems' "real" cost growth consistently exceeded annual changes in DoD deflator values, even when SARs listed portions of space systems program budget economic-based changes that were directly tied to updates to annual DoD deflators.

One expert we interviewed opined that it is not the role of an inflation index to control contractor costs or incentivize contractors to implement productivity improvements. Rather, tools such as DCMA negotiations of contractors' FPRA and acquisition strategies advocating for increased contractual competition are a few examples of potential "best business practices" for implementing cost controls. Since budget

---

[36] Although the intent of the Steel Vessel Index is to better capture the costs of shipbuilding than the GDP deflator does, Keating et al., 2008, suggest that the Steel Vessel Index does not accurately cover the materials used in building a modern ship, i.e., the costs of iron and steel are overrepresented in the index.

indexes simply reflect the consequence or outcomes of using these policy tools, this same expert concluded that "Budget growth is a symptom of a problem, not a cause."

Suggested Actions

We recommend that the following three actions be considered:

1. Today, Global Insight indexes are used by NRO and can, under current SMC guidance, be used across the defense space cost government community as the primary basis for computing weighted composite space inflation index values. Even if this approach turns out to be the best way to ensure accurate TY dollar budgets for major space system defense acquisition programs, *we suggest that OSD CAPE conduct systematic reviews at both the services' cost agency and acquisition command levels to validate the accuracy of space-unique inflation indexes.*

    A recent SMC/Commander cost-escalation/inflation guidance memo allows program offices to use inflation rates other than the OSD published ones, but they are required to provide the rationale for using different rates and the estimated effect of using them.[37] The program offices must request an exemption from SAF/Financial Management to use these unique rates in generating their program office estimates.[38] Even though this is guidance, the Air Force space cost community has yet to validate the accuracy of Global Insight's projected indexes.

2. Given the significant portion of contractor direct labor costs for space systems contracts, *we also recommend that OSD AT&L, in coordination with the appropriate offices within OSD CAPE and DCMA, initiate an in-depth defense space industrial base assessment of the rationale for the differences in current FRPA and projected direct labor rates over the next three to ten years.* This assessment should cut across the prime contractors and major subcontractors based on demographics, business base changes, and other significant cost drivers.

3. Finally, *we recommend that government space system PMs be encouraged to use an economic price adjustment table within requests for proposals as part of pricing instructions for contractors that estimate space system contract FFP costs for procuring follow-on satellites (e.g., AEHF SV-5 and SV-6) and setting prices for other contract line item numbered deliverables.* EPA values should be set at or below negotiated FPRA costs to motivate contractor PMs to execute cost controls for managing direct labor pool skill mix and, as necessary, to reduce the "standing army" of staff while other engineers are mitigating risks.

---

[37]   SMC, 2011.

[38]   AFI 65-502, 1994.

Bibliography

AFI—*See* Air Force Instruction.

"AEHF-1 Arrives at Its Operational Orbit 14-Months Journey," *Air Force Print News Today,* October 25, 2011.

Aerospace Corporation, meeting with authors on space acquisition issues, April 17, 2013.

AFSPC—*See* Air Force Space Command.

"Air Force Commander: GPS III, OCX Delayed," *Inside GNSS*, May 28, 2012. As of July 18, 2013: http://www.insidegnss.com/node/3096

Air Force Instruction 65-502, "USAF Instruction 65-502 Financial Management Inflation," OPR: SAF/FMCEE, January 21, 1994.

Air Force Space Command, "Resiliency and Disaggregated Space Architectures: White Paper," August 2013.

Air Force Space Command Launch, Ranges, and Networks Division, "Launch Information Support Network (LISN)," undated, Not available to the general public. As of July 19, 2013: https://lisn.peterson.af.mil

Alic, John, et al., *Beyond Spinoff: Military and Commercial Technologies in a Changing World*, Boston: Harvard University Press, 1992.

BEA—*See* Bureau of Economic Analysis.

Blickstein, Irv, Michael Boito, Jeffrey A. Drezner, James Dryden, Kenneth Horn, James G. Kallimani, Martin C. Libicki, Megan McKernan, Roger C. Molander, Charles Nemfakos, Chad J.R. Ohlandt, Caroline Reilly, Rena Rudavsky, Jerry Sollinger, Katharine Watkins Webb, and Carolyn Wong, *Root Cause Analyses of Nunn-McCurdy Breaches, Volume 1:* Zumwalt-*Class Destroyer, Joint Strike Fighter, Longbow Apache, and Wideband Global Satellite*, Santa Monica, Calif.: RAND Corporation, MG-1171/1-OSD, 2011. As of December 3, 2014: http://www.rand.org/pubs/monographs/MG1171z1.html

Boeing—*See* Boeing Space & Intelligence Systems.

Boeing Space & Intelligence Systems, meeting with authors on space acquisition issues, April 18, 2013.

Bohn, Michal, Eric Mosier, Wayne Salis, James Smirnoff, Kevin Connor, and Nicole Hamaker, *Inflation: Lessons Learned*, National Reconnaissance Office briefing to Society for Cost Estimating and Analysis, June 2012.

Botwin, Brad, and Christopher Nelson, "U.S. Space Industrial 'Deep Dive': Final Dataset Findings," U.S. Space Industry Deep Dive Assessment, May 2013. As of October, 20, 2013: http://www.bis.doc.gov/index.php/space-deep-dive-results

Brinton, Turner, "Pentagon Cancels T-Sat Program, Trims Missile Defense," *Space News*, April 6, 2009.

Bureau of Economic Analysis, *BLS Handbook of Methods,* Chapter 14, "Producer Prices," undated (a). As of December 3, 2014:
http://www.bls.gov/opub/hom/homch14.htm

———, *Interactive Tables*, U.S. Department of Commerce, undated (b). As of December 3, 2014:
http://www.bea.gov/iTable/index.cfm

———, "National Data: National Income and Product Account Tables," U.S. Department of Commerce, undated (c). As of January 7, 2015:
http://www.bea.gov/iTable/iTable.cfm?ReqID=9&step=1#reqid=9&step=3&isuri=1&903=4

Bureau of Labor Statistics, *Databases, Tables and Calculators by Subject*, undated. As of December 3, 2014:
http://www.bls.gov/data/#prices

Butler, Amy, "Pentagon Still Relying on Spot Buys for Milsatcom," *Aviation Week & Space Technology*, March 18, 2013. As of November 13, 2013:
http://www.aviationweek.com/Article.aspx?id=/article-xml/AW_03_18_2013_p56-557815.xml

CAPE—*See* Office of the Secretary of Defense, Cost Assessment and Program Evaluation.

Carter, Ashton, Under Secretary of Defense for Acquisition, Technology, and Logistics, "Establishment of the Annual Global Positioning System (GPS) Enterprise Review (AGER)," Acquisition Decision Memorandum, November 23, 2009.

Chokshi, Niraj, "Education Costs Rising Faster Than Health Care," *The Atlantic*, August 24, 2009. As of December 3, 2014:
http://www.theatlantic.com/business/archive/2009/08/
education-costs-rising-faster-than-health-care/23705/

Congressional Budget Office, "A Comparison of Science and Technology Funding for DoD's Space and Nonspace Programs," January 15, 2008.

———, *Baseline Economic Forecast—February 2013 Baseline Projections*, February 5, 2013. As of December 3, 2014:
http://www.cbo.gov/publication/43902

———, "Baseline Projections—Real GDP," January 8, 2014. As of December 4, 2014:
http://www.quandl.com/CBO/PROJ_REALGDP-Baseline-Projections-Real-GDP

Connor, Kathryn, and James Dryden, *New Approaches to Defense Inflation and Discounting*, Santa Monica, Calif.: RAND Corporation, RR-237-OSD, 2013. As of December 3, 2014:
http://www.rand.org/pubs/research_reports/RR237.html

Cooley, Bill, "Global Positioning Systems Directorate: GPS Program Update to Civil GPS Service Interface Committee (CGSIC)," September 17, 2013. As of January 28, 2014:
http://www.gps.gov/cgsic/meetings/2013/cooley.pdf

Coonce, Tom, Bob Bitten, Joe Hamaker, and Henry Hertzfeld, "NASA Productivity Study," December 24, 2008. As of October 20, 2013:
http://www.gwu.edu/~spi/assets/docs/Productivity.pdf

Davis, Dee Ann, "Air Force Plans Shift to Fixed-Price Contract for GPS III," *Inside GNSS News*, March 19, 2012. As of November 24, 2013:
http://www.insidegnss.com/node/2982

———, "Air Force Examining Broader Options for Next GPS III Satellite Buy," *Inside GNSS,* April 30, 2013. As of November 2, 2013:
http://www.insidegnss.com/node/3538

DCMA—*See* Defense Contract Management Agency.

Defense Advanced Research Projects Agency, "System F6," undated. As of May 22, 2014:
http://www.darpa.mil/Our_Work/TTO/Programs/System_F6.aspx

Defense Contract Management Agency, *Forward Pricing Rates,* undated. As of September 24, 2013:
http://guidebook.dcma.mil/41/index.cfm

"Defense Support Program Satellite Decommissioned," *GlobeNewswire,* July 31, 2008. As of November 8, 2013:
http://globenewswire.com/news-release/2008/07/31/382292/147496/en/Defense-Support-Program-Satellite-Decommissioned.html

DoC—*See* U.S. Department of Commerce.

Dopkeen, Bess, and Jon Sweet, "Evolutionary Acquisition for Space Efficiency (EASE): Briefing for the Cost Community–DoDCAS," February 17, 2011. Not available to the general public.

DOT&E—*See* Office of the Director, Operational Test and Evaluation.

Ferster, Warren, "DARPA Cancels Formation-Flying Satellite Demo," *Space News,* May 17, 2013. As of November 26, 2013:
http://www.spacenews.com/article/military-space/35375darpa-cancels-formation-flying-satellite-demo

Foust, Jeff, "An Opening Door for Hosted Payloads," *The Space Review,* October 29, 2012. As of December 23, 2014:
http://www.thespacereview.com/article/2179/1

Gansler, Jacques S., Under Secretary of Defense for Acquisition and Technology, "Acquisition Decision Memorandum for Advanced Extremely High Frequency (AEHF) Program," May 26, 2000.

GAO—*See* U.S. Government Accountability Office.

Gruss, Mike, "WGS Launch on Delta 4 Now Slated for May," *Space News,* March 29, 2013. As of November 13, 2013:
http://www.spacenews.com/article/military-space/34620wgs-launch-on-delta-4-now-slated-for-may

———, "U.S. Air Force Claims Big Savings on EELV Block Buy," *Space News,* January 31, 2014. As of May 22, 2014:
http://www.spacenews.com/article/military-space/39348us-air-force-claims-big-savings-on-eelv-block-buy

Hogan, Greg, "Estimating Inflation for Space Programs" Air Force Cost Analysis Agency, Joint Air Force and NRO briefing provided to RAND, October 24, 2013.

Horowitz, Stanley A., Alexander O. Gallo, Daniel B. Levine, Robert J. Shue, and Robert W. Thomas, *The Use of Inflation Indexes in the Department of Defense,* Alexandria, Va.: Institute for Defense Analyses, P-4707, May 2012.

———, *The Use of Inflation Indexes in the Department of Defense,* Alexandria, Va.: Institute for Defense Analyses, Proceedings from the Tenth Annual Acquisition Research Symposium Cost Estimating, April 1, 2013.

Host, Pat, "Air Force Awards Raytheon $70 Million Contract for FAB-T Ground Terminals," *Defense Daily,* September 14, 2012.

Hosted Payload Alliance, *Hosted Payload Discussion*, briefing, Santa Monica, Calif.: RAND Corporation, December 7, 2012.

Hura, Myron, Gary McLeod, Lara Schmidt, Manual Cohen, Mel Eisman, and Elliot Axelband, "Space Capabilities Development: Implications of Past and Current Efforts for Future Programs," Santa Monica, Calif.: RAND Corporation, September 2007. Not available to the general public.

Hura, Myron, Manuel Cohen, Elliot Axelband, Richard Mason, and Mel Eisman, "Space Capabilities Development: Continuing Difficulties and Suggested Actions," Santa Monica, Calif.: RAND Corporation, June 2011. Not available to the general public.

"Inflation Comparisons," NRO briefing, August 2009.

Inforum, "Lift Model," undated. As of June 16, 2014:
http://www.inforum.umd.edu/services/models/lift.html

Ingols, Cynthia, et al., "Implementing Acquisition Reform: A Case Study on the Joint Direct Attack Munitions," Defense Systems College Management, Ft. Belvoir, Va., July 1998. As of September 5, 2014:
http://www.acquisition.gov/sevensteps/library/jdamsuccess.pdf

Jackson, Keoki, "GPS Modernization: GPS III on the Road to the Future," Stanford PNT Symposium, November 13–14, 2012. As of November 9, 2013:
http://scpnt.stanford.edu/pnt/PNT12/2012_presentation_files/14-Jackson_presentation.pdf

Keating, Edward G., Robert Murphy, John F. Schank, and John Birkler, *Using the Steel-Vessel Material-Cost Index to Mitigate Shipbuilder Risk*, Santa Monica, Calif.: RAND Corporation, TR-520-NAVY, 2008. As of December 3, 2014:
http://www.rand.org/pubs/technical_reports/TR520.html

"Lockheed Martin-Built DSCS Satellites Achieve Historic Milestone," *PR Newswire*, February 17, 2009. As of November 15, 2013:
http://www.prnewswire.com/news-releases/lockheed-martin-built-dscs-satellites-achieve-historic-milestone-65741007.html

Lockheed Martin, "Global Positioning System (GPS)," undated. As of November 2, 2013:
http://www.lockheedmartin.com/us/products/gps.html

———, "Lockheed Martin Completes Navigation Payload Milestone for GPS III Prototype," May 29, 2011. As of November 2, 2013:
http://www.lockheedmartin.com/us/news/press-releases/2012/may/0529-ss-gpsIII.html

———, "U.S. Air Force Awards Lockheed Martin Contract for Third and Fourth GPS III Satellites," January 2012a. As of December 1, 2014:
http://www.lockheedmartin.com/us/news/press-releases/2012/january/0112_ss_gps.html

———, *Lockheed Martin Completes On Orbit Testing of Second AEHF Satellite*, November 2012b.

———, meeting with authors on space acquisition issues, April 16, 2013a.

———, "Lockheed-Martin Built SBIRS GEO-2 Missile Defense Early Warning Satellite Certified for Operation," December 17, 2013b. As of November 25, 2014:
http://www.lockheedmartin.com/us/news/press-releases/2013/december/1217-ss-sbirs-geo-2.html

———, "U.S. Air Force Awards Lockheed Martin Contract for Next Two SBIRS Missile Defense Early Warning Satellites," Sunnyvale, Calif., June 24, 2014. As of August 29, 2014:
http://www.lockheedmartin.com/us/news/press-releases/2014/june/0624-ss-sbirs.html

Maryland Interindustry Forecasting Project, Research Memorandum No. 20, undated. As of February 16, 2015:
http://www.inforum.umd.edu/papers/wp/wpa/1969/mifp20.pdf

McCullough, Amy, "Space Acquisition with EASE," *Air Force Magazine*, February 16, 2011. As of November 24, 2011:
http://www.airforcemag.com/Features/modernization/Pages/box021611ease.aspx

McKinsey & Company, meeting with authors on space acquisition study, December 10, 2012.

McNew, Gregory J., *An Examination of the Patterns of Failure in Defense Acquisition Programs*, Chapter 2: "Literature Review," February 2011.

Mehta, Aaron, "SpaceX Wins U.S. Air Force EELV Missions," *Defense News*, December 6, 2012. As of November 24, 2013:
http://www.defensenews.com/article/20121206/DEFREG02/312060006

Moody, Jay A., *Achieving Affordable Operational Requirements on the Space Based Infrared System (SBIRS) Program: A Model for Warfighter and Acquisition Success?* paper presented to the Research Department Air Command and Staff College, AU/ACSC/97-0548, March 2007. As of January 28, 2014:
http://www.dtic.mil/get-tr-doc/pdf?AD=ADA397934

National Coordination Office for Space-Based Positioning, Navigation, and Timing, "Space Segment," updated September 17, 2013. As of September 19, 2013:
http://www.gps.gov/systems/gps/space/

———, "Space Segment," updated December 12, 2014. As of December 23, 2014:
http://www.gps.gov/systems/gps/space/

National Reconnaissance Office, *NSCM Inflation Study v2—Findings for Discussion*, January 2004.

———, *Inflation Comparisons*, briefing, August 2009.

———, "Inflation: Lessons Learned," NRO SCEA briefing, June 2012.

National Security Space Strategy: Unclassified Summary, ODNI, DoD, January 2011.

Northrop Grumman, meeting with authors on space acquisition issues, June 6, 2013a.

———, "Facts: Advanced EHF Protected Satellite Communication Payloads," September 6, 2013b. As of November 17, 2013:
http://www.northropgrumman.com/Capabilities/AdvancedEHFPayloads/Documents/pageDocs/AEHF_fact_sheet.pdf

NRO—*See* National Reconnaissance Office.

Office of Management and Budget, "Historical Tables," undated. As of December 3, 2014:
http://www.whitehouse.gov/omb/budget/Historicals/

Office of the Assistant Secretary of the Air Force for Acquisition (Space) (SAF/AQS), discussions with staff about space acquisition and the status of the GPS III program, January 2013.

Office of the Director, Operational Test and Evaluation (DOT&E), DOT&E FY 2002 Annual Report, "Wideband Gapfiller Satellite," undated (a), pp. 307–308.

———, DOT&E FY 1999 Annual Report, "Space-Based Infrared System (SBIRS)," undated (b), pp. 159–166. As of May 22, 2014:
http://www.dote.osd.mil/pub/reports/FY1999/

Office of the Secretary of Defense, "Directive-Type Memorandum (DTM) 09-027—Implementation of the Weapon Systems Acquisition Reform Act of 2009," December 4, 2009, Incorporating Change 4, January 11, 2013.

Office of the Secretary of Defense, Cost Assessment and Program Evaluation (CAPE), "FY 2011 Annual Report of Cost Assessment Activities," Washington, D.C.: Department of Defense, February 2012.

Office of the Under Secretary of Defense for Acquisition, Technology, and Logistics, *Report of the Defense Science Board/Air Force Scientific Advisory Board Joint Task Force on Acquisition of National Security Space Programs,* May 2003.

———, Memorandum on "Better Buying Power 2.0: Continuing the Pursuit for Greater Efficiency and Productivity in Defense Spending," November 2012.

———, Memorandum on "Implementation Directive for Better Buying Power 2.0—Achieving Greater Efficiency and Productivity in Defense Spending," April 2013a.

———, "Performance of the Defense Acquisition Systems, 2013 Annual Report," 2013b.

Office of the Under Secretary of Defense (Comptroller), "National Defense Budget Estimates for FY 2015," Washington, D.C.: Department of Defense, April 2014.

OUSD (AT&L)—*See* Office of the Under Secretary of Defense for Acquisition, Technology, and Logistics.

OUSD (C)—*See* Office of the Under Secretary of Defense, Comptroller.

"Parts Testing Drives Up GPS III Program Costs, Forces Prime to Forego $70 Million Incentive Fee," *Inside GNSS*, April 21, 2012, As of October 28, 2013:
http://www.insidegnss.com/node/3033

Pawlikowski, Elllen M., "Moving Beyond Acquisition Reform's Legacy," *Air Force Space Command High Frontier*, Vol. 2, No. 2, 2006, pp. 27–32.

———, "SMC Cost Guidance—Cost Escalation/Inflation for SMC Cost Estimates Memorandum for All Staff and Directorates," Los Angles, Calif.: Department of the Air Force Headquarters, Space and Missile Systems Center, November 23, 2011.

Pawlikowski, Ellen, Doug Loverro, and Tom Cristler, "Space: Disruptive Challenges, New Opportunities, and New Strategies," *Strategic Studies Quarterly*, Spring 2012, pp. 27–54.

Peck, Michael, "Air Force Looking at GPS III Production Alternatives," *Defense News*, July 7, 2014. As of August 29, 2014:
http://www.c4isrnet.com/apps/pbcs.dll/article?AID=2014307070002

Perry, William, Secretary of Defense, "Specifications and Standards—A New Way of Doing Business," Memorandum for Secretaries of the Military Departments, June 29, 1994.

Public Law 111-23, *Weapon System Acquisition Reform Act (WSARA) of 2009*, Title I, Section 101, May 22, 2009.

Rubin, Donald. "For Objective Causal Inference, Design Trumps Analysis," *The Annals of Applied Statistics*, 2008, Vol. 2, No. 3, pp. 808–840.

SAF/AQS—*See* Office of the Assistant Secretary of the Air Force for Acquisition (Space).

SBIRS SAR—*See* Space-Based Infrared System Selected Acquisition Report.

Sega, Ronald M., DoD Executive Agent for Space, DoD Memorandum, Subject: "Back to Basics and Implementing a Block Approach for Space Acquisition," March 14, 2007.

Shalal-Esa, Andrea, "U.S. Still Probing Security Satellite Failure," Reuters, January 6, 2009. As of November 8, 2013:
http://www.reuters.com/article/2009/01/06/us-northrop-satellite-idUSTRE5055DW20090106

Simpson, Jason, "Canty: OCX Work Would Not Have Started Yet if Tied to Space Segment," *Inside the Air Force*, April 18, 2008a.

———, "Officials Question Worth of Splitting Ground and Space Contracts," *Inside the Air Force*, December 26, 2008b.

Sirak, Michael, "Air Force Aims to Transform Space Acquisition Process," *Satellite TODAY News Feed*, February 27, 2006. As of November 24, 2013:
http://www.satellitetoday.com/publications/st/feature/2006/02/27/
air-force-aims-to-transform-space-acquisition-process/

SMC—*See* Space and Missile Systems Center.

Space and Missile Systems Center, *Systems Engineering Primer and Handbook, Concepts Processes and Techniques*, January 15, 2004.

———, Cost Guidance, November 2011.

———, "Program Operating Plan (POP), Version 10.3, AFSPC, Version 10.3, September 28, 2012. Not available to the general public.

———, "USCM Inflation Index Research OSD vs. NRO for Historical Data Normalization," briefing, March 2013a.

———, meeting with authors on space acquisition issues, El Segundo, Calif., June 28, 2013b.

Space and Missile Systems Center/MILSATCOM Joint Program Office/Advanced EHF Program Office, *Advanced Extremely High Frequency Satellite Communications System Test and Evaluation Master Plan*, December 15, 2000. Not available to the general public.

Space News staff, "Lockheed Receives $42M for Long-Lead SBIRS Parts," *Space News*, September 30, 2013. As of January 20, 2014:
http://www.spacenews.com/article/military-space/37461lockheed-receives-42m-for-long-lead-sbirs-parts

Starosta, Gabe, "GPS OCX Development Challenged by Cybersecurity Requirements," *Inside the Air Force,* March 30, 2012.

The Tauri Group, "U.S. Industrial Base Analysis for Space Systems," presentation at the Defense Manufacturing Conference, Anaheim, Calif., November 29, 2011.

Tirpak, John, "Challenges Ahead for Military Space," *Air Force Magazine*, January 2003. As of November 16, 2013:
http://www.airforcemag.com/MagazineArchive/Pages/2003/January%202003/0103space.aspx

Trauberman, Jeff, "Space Cost Reduction: Wideband Global SATCOM (WGS) Case Study," Aerospace Industries Association–Space Council, May 29–30, 2013.

"Two Years Later, SBIRS Geo-1 Finally Declared Operational," *Space News*, June 7, 2013. As of January 20, 2014:
http://www.spacenews.com/article/military-space/35683two-years-later-sbirs-geo-1-finally-declared-operational

U.S. Air Force, *Military Satellite Communication Space Modernization Initiative Investment Plan*, report to Congressional Committees, April 2012a.

————, *Lower Cost Solutions for Providing Global Positioning System Capability*, Report to Congressional Committees, August 15, 2012b.

U.S. Air Force Cost Analyses Agency (AFCAA), *Estimating Inflation for Space Programs*, October 2013.

U.S. Air Force SMC/CC Memorandum, *Cost Guidance—Cost Escalation/Inflation for SMC Cost Estimates*, November 23, 2011.

U.S. Air Force SMC/FM, *USCM Inflation Index Research: OSD vs. NRO for Historical Data Normalization*, SMC Cost IPT Tecolote briefing, March 2013.

U.S. Air Force System Metrics and Reporting Tool (SMART) Database, "Monthly Acquisition Reports: AEHF," Hanscom AFB, September 2005–November 2012. Not available to the general public.

————, "Monthly Acquisition Reports: GPS Block IIF," Hanscom AFB, February 2005–September 2012. Not available to the general public.

————, "Monthly Acquisition Reports: GPS III," Hanscom AFB, May 2005–September 2012. Not available to the general public.

————, "Monthly Acquisition Reports: SBIRS," Hanscom AFB, July 2009–September 2012. Not available to the general public.

————, "Monthly Acquisition Reports: WGS," Hanscom AFB, April 2009–July 2012. Not available to the general public.

U.S.-China Economic and Security Review Commission, "2011 Report to Congress of the U.S.-China Economic and Security Review Commission," One Hundred Twelfth Congress, First Session, November 2011. As of December 4, 2014:
http://origin.www.uscc.gov/sites/default/files/annual_reports/annual_report_full_11.pdf

U.S. Department of Commerce, Bureaus of Industry and Security, "U.S. Space Industry 'Deep Dive'," presented at the National Defense Industrial Association National Security Space Policy and Architecture Symposium, June 25–26, 2013, Chantilly, Va.

U.S. Department of Defense, "National Defense Budget Estimates for FY 2013," Washington, D.C., March 2012.

————, "Selected Acquisition Reports: Advanced Extremely High Frequency Satellite Communications System," December 2001 through December 2012.

————, "Selected Acquisition Reports: Evolved Expendable Launch Vehicle," December 2007 through December 2012.

————, "Selected Acquisition Reports: GPS-IIIA," June 2008 through December 2012.

————, "Selected Acquisition Reports: NAVSTAR GPS," December 1996 through December 2012.

————, "Selected Acquisition Reports: Space Based Infrared System," December 1995 through December 2012.

———— "Selected Acquisition Reports: Wideband Gapfiller System," December 2001 through December 2006.

————, "Selected Acquisition Reports: Wideband Global SATCOM," December 2007 through December 2012.

U.S. Government Accountability Office, "Defense Satellite Communications: Alternative to DoD's Satellite Replacement Plan Would Be Less Costly," GAO/NSIAD-97-159, July 1997.

————, "Military Space Operations: Common Problems and Their Effects on Satellite and Related Acquisitions," GAO-03-825R, June 2003a.

————, "Despite Restructuring, SBIRS High Program Remains at Risk of Cost and Schedule Overruns," GAO-04-48, October 2003b.

————, "Space Acquisitions: DoD Needs to Take More Action to Address Unrealistic Initial Cost of Space Systems," GAO-07-96, November 2006.

————, "Defense Acquisitions: Assessments of Selected Weapon Programs," GAO-07-406SP, March 2007a.

————, "Space Acquisitions: Actions Needed to Expand and Sustain Use of Best Practices," GAO-07-730T, April 19, 2007b.

————, "Officials Question Worth of Splitting Ground and Space Contracts," *Inside the Air Force*, December 26, 2008.

————, "Defense Acquisitions: Assessments of Selected Weapon Programs," GAO-09-326-SP, March 2009a.

————, "Global Positioning System: Significant Challenges in Sustaining and Upgrading Widely Used Capabilities," GAO-09-325, April 2009b.

————, "Space Acquisitions: Government and Industry Partners Face Substantial Challenges in Developing New DoD Space Systems," GAO 09-648T, April 30, 2009c.

————, "Defense Acquisitions: Challenges in Aligning Space System Components," GAO-10-55, October 2009d.

————, "GPS: Challenges in Upgrading and Sustaining Capabilities Persist," GAO-10-636, September 2010.

————, "Space and Missile Defense Acquisitions: Periodic Assessment Needed to Correct Parts Quality Problems in Major Programs," GAO-11-404, June 2011.

————, "Space Acquisitions: DOD Faces Challenges in Fully Realizing Benefits of Satellite Acquisition Improvements," GAO-12-563T, March 21, 2012.

————, "Assessments of Selected Weapon Programs," GAO-13-294SP, March 28, 2013a.

————, "Space Acquisitions: DOD Is Overcoming Long-Standing Problems, but Faces Challenges to Ensuring Its Investments Are Optimized," GAO-13-508T, April 2013b.

————, "Global Positioning System: A Comprehensive Assessment of Potential Options and Related Costs Is Needed," GAO-13-729, September 2013c.

————, "Space Acquisition Issues in 2013," *Air & Space Power Journal*, September–October 2013d, pp. 11–28.

USCM Inflation Index Research, "OSD vs. NRO for Historical Data Normalization," Los Angeles, Calif.: USAF Space and Missile Systems Center, March 2013.

Welsh, Brian, *Inflation Research and Treatment*, TASC, USAF SMC Cost IPT, June 17–18, 2009.

White, Christopher P., *An Analysis of Total System Performance Responsibility in Air Force Acquisitions*, Thesis, Air Force Institute of Technology, AFIT/GAQ/ENS/01M-04, 2001. As of November 23, 2013:
http://www.dtic.mil/dtic/tr/fulltext/u2/a391226.pdf

Whitley, Gigi, "Air Force Officials Mull Accelerating Launch of Advanced EHF Satellite," *Inside the Air Force*, May 7, 1999.

Worthen, Ben, "Rising Computer Prices Buck the Trend," *The Wall Street Journal*, December 13, 2010. As of December 3, 2014:
http://online.wsj.com/article/SB10001424052748704681804576017883787191962.html

Younossi, Obaid, Mark V. Arena, Robert S. Leonard, Charles Robert Roll, Jr., Arvind Jain, and Jerry M. Sollinger, *Is Weapon System Cost Growth Increasing? A Quantitative Assessment of Completed and Ongoing Programs,* Santa Monica, Calif.: RAND Corporation, MG-588-AF, 2007. As of December 12, 2014:
http://www.rand.org/pubs/monographs/MG588.html

Younossi, Obaid, Mark A. Lorell, Kevin Brancato, Cynthia R. Cook, Mel Eisman, Bernard Fox, John C. Graser, Yool Kim, Robert S. Leonard, Shari Lawrence Pfleeger, and Jerry M. Sollinger, *Improving the Cost Estimation of Space Systems: Past Lessons and Future Recommendations,* Santa Monica, Calif.: RAND Corporation, MG-690-AF, 2008. As of December 22, 2014:
http://www.rand.org/pubs/monographs/MG690.html